EVERYTHING ABOUT ASTRONAUTS

EVERYTHING ABOUT
ASTRONAUTS

FASCINATING FUN FACTS AND TRIVIA ABOUT
ASTRONAUTS FOR TEENS AND ADULTS

by
MARIANNE JENNINGS

VOL
I

KNB

DEDICATION

For Nome, the North Star in my sky.

&

To the men and women who have gone before
and those who will go forward to show us that the
impossible really is possible.

Copyright and Disclaimer

Everything About Astronauts Volume 1:
Fascinating Fun Facts and Trivia about Astronauts for Teens and Adults

Knowledge Nuggets Series by Marianne Jennings

Edited by Karina Hamalainen, Joe Levit, and Skye Loyd.
Fact checked by Hank Musolf
Cover design: Matt Davies
Interior design and formatting: Bea Reis Custodio

While all attempts have been made to verify the information provided in this publication, neither the author nor the publisher assumes any responsibility for errors, omissions, or contrary interpretations on the subject matter herein. This book is for entertainment purposes only. The views expressed are those of the author alone, and should not be taken as expert instruction or commands. The reader is responsible for his or her own actions.

Library of Congress Control Number: 2021922477
ISBN 978-1-7342456-4-6 (paperback)
ISBN 978-1-7342456-5-3 (ebook)

Trademarks that are mentioned are done without written consent and can in no way be considered an endorsement from the trademark holder.

Disclaimer: No astronauts were harmed during the making of this book.

CONTENTS

HOW TO READ THIS BOOK

This book is divided into topics, and the facts are in bite-size nuggets. There is no need to read this book cover to cover. Just pick a subject that seems interesting and jump right in!

Glossary

There are several acronyms and a few technical terms throughout this book. Please refer to the glossary found at the end of the book to find definitions and short explanations.

Index

If there's a particular topic or person you'd like to learn more about, please refer to the index at the very back of the book.

Quiz Yourself

To test yourself and your friends with what you've learned, you'll find a fun, short quiz with answers in the back.

Please Bookmark the RESOURCES Page

so you can easily access all the videos, images and other resources mentioned throughout this book with direct links to each.

KnowledgeNuggetBooks.com/resources

SPECIAL BONUS

As a **special bonus** and as a **thank you** for purchasing this book, I'm giving you **early access and a free download of the first chapter** of the upcoming *Everything About Astronauts Volume 2*.

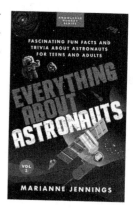

Volume 2 covers intriguing topics about what it's like to live and work in space, what foods astronauts can and can't eat, random space records, weird space injuries and so much more.

It's all FREE.

Download your bonus ebook chapter of *Everything About Astronauts Volume 2* here:

http://bit.ly/astronaut-bonus-chapter

or

SCAN ME

Enjoy!

INTRODUCTION

"Space is for everybody. It's not just for a few people in science or math, or for a select group of astronauts. That's our new frontier out there, and it's everybody's business to know about space."
— **Christa McAuliffe, *Challenge* astronaut, the first teacher selected to fly into space.**

On July 20, 1969, the world was glued to their televisions as they watched astronauts Neil Armstrong and Buzz Aldrin take the first steps on the Moon. Humans traveling to space has been a reality and a part of our lives since 1961. For many of us, we have never lived in a world where we have known anything different.

Humans have accomplished what was once thought impossible. Humans have traveled to space and back, walked on the Moon and returned home safely and are now living in space for months at a time. We now look forward to the days in a not-so-distant future where a woman will step foot on the Moon, and humans will travel to Mars. These are exciting and adventurous times. We continue to prove the impossible is actually possible.

As I write this, NASA's Mars Rover *Perseverance* recently landed on the surface of Mars successfully and has been sending back pictures, videos and even the sounds of Mars. We are getting to see what the surface of another planet looks and sounds like, and not only that, the pictures are high res and in color. Do you realize how far and fast that information has to travel through space to reach us back on Earth? It's mind-bending, and yet we don't even give it a second thought.

I've long been fascinated by all things space and about the men and women who have been able to travel and explore space. I wanted to write a book that appeals to people of all ages who share that same fascination. We are entering a time where there is renewed interest in space exploration and the men and women who get to be our eyes and ears exploring it.

This book was originally one very long book, but to make it easier to digest, it has been broken up into two volumes. Volume 1 covers mostly astronaut topics that are about famous astronauts and about things that take place "on the ground," where Volume 2 covers astronaut life "up in space."

I hope you enjoy reading, and will find fun facts, stories and astronaut trivia that will excite you, cause you to wonder and make you say, "Wow, I didn't know that!"

Never stop learning.

Marianne

ASTRONAUT, COSMONAUT AND OTHER 'NAUTS

WHAT'S A 'NAUT?

Each country has a different name for people who go to space. But most end in "naut" which comes from the Greek word "nautes," which means "sailor."

ASTRO + NAUT = STAR SAILOR

The "astro" part of astronaut comes from the Greek word "astron," meaning "star." Astronauts most often refer to people from the United States, Canada, Europe and Japan who travel into space.[1] NASA scientists thought they were the first to coin the word astronaut in the 1950s, but science fiction has used the term since the 1920s.[2]

COSMO + NAUT = UNIVERSE SAILOR

Russia—and the Soviet Union before that—use "cosmo" as a prefix, which is the Greek word for "universe." So the literal translation for cosmonaut is "universe sailor."[3]

TAIKO + NAUT = SPACE SAILOR OR TRAVELER OF THE UNIVERSE

The Chinese officially use the name "yuhangyuan," (pronounced you-hang-you-en) which means "traveler of the Universe." For those who don't speak Chinese, we use the name "taikonauts." This comes from the Chinese word, "taikong," which means "space" or "cosmos." Pair that with "naut" and it means "space sailor."[4]

SPATIO + NAUT = SPACE SAILOR

The term "spationaut" refers to a French astronaut or is used as a casual term for a European astronaut from the European Space

Agency (ESA). In Latin, "spatium" means "space." Add that to "naut," and that equals "space sailor."

VYOMA + NAUT = SKY OR SPACE SAILOR

Men and women from India who travel into space are known as "vyomanaut" (pronounced VEE-oh-mo-nauts) which uses the Sanskrit word "vyoma," which means sky or space.[5] Again, the Greek word for sailor is the preferred suffix and translates into sky or space sailor.

MARSONAUTS

An experiment called Mars500 ran a full-length, simulated human mission to Mars. In June 2010, six men referred to as "marsonauts" were locked into an Earth-based-spaceship near Moscow, Russia, for 520 days. This is the amount of time it would take to fly to Mars and back, including 30 days exploring the surface of Mars.[6]

TERRANAUTS

Terra is Latin for "land." Some volunteers spend at least three months on bedrest in experiments to simulate some aspects of the experience astronauts have orbiting Earth. These volunteers have been nicknamed terranauts.[7]

NOTES

1. John M. Logsdon, "Astronaut," Encyclopedia Britannica, February 5, 2020, https://www.britannica.com/topic/astronaut.

2. Colin Burgess, *Selecting the Mercury Seven the Search for America's First Astronauts* (New York: Springer, 2011), 30.

3. "Astronaut," Vocabulary.com, accessed May 11, 2021, https://www.vocabulary.com/dictionary/astronaut.

4. "Taikonaut," Oxfordreference.com, accessed May 11, 2021, https://doi.org/10.1093/oi/authority.20110803101916587.

5. Anil Ananthaswamy, "Wanted: Four 'Vyomanauts' for Indian Spaceflight," *New Scientist* (1971), January 5, 2010, https://www.newscientist.com/article/dn18338-wanted-four-vyomanauts-for-indian-spaceflight/.

6. "Mars Confinement Tips," European Space Agency, April 22, 2020, http://www.esa.int/ESA_Multimedia/Videos/2020/04/Mars_confinement_tips.

7. Mary Roach, *Packing for Mars: The Curious Science of Life in the Void* (New York: WW Norton, 2011), 214.

BEEN THERE, DONE THAT

Serious Bragging Rights

HOW MANY PEOPLE HAVE BEEN TO SPACE?

Fewer than 600 people who have gone into space. Space begins 62 miles, or 100km, above the surface of the Earth at the Kármán line, according to the International Astronautical Federation. Since Yuri Gagarin became the first human to launch into orbit on April 12, 1961, a total of 584 people from 41 different nations have gone into space above the Kármán line as of October 2021.[1]

MALE VS. FEMALE ASTRONAUTS

Of those 584 individuals who have been to space, 515 were men and 69[2] were women.[3]

TO THE MOON AND BACK

There are only 24 men who can say that they've been to the Moon. This includes not only the 12 who walked on the Moon, but also the other 12 who left Earth's orbit, and then circled and orbited the Moon.

HOW MANY PEOPLE HAVE BEEN TO THE INTERNATIONAL SPACE STATION?

The International Space Station (ISS) orbits the Earth at an average altitude of 250 miles (402 km). As of April 2021, over 244 people from 19 different countries have been to the ISS.[4]

TOP OF THE WORLD AND LAUNCHED INTO SPACE

Only two people have been to space *and* summitted Mt. Everest. In 2009, American astronaut Scott Parazynski became the first person to do both.[5] Italian astronaut Maurizio Cheli became the second person when he summited Everest in 2018.[6]

WHO HAS SEEN BOTH THE BOTTOM OF THE SEA AND THE EARTH FROM SPACE?

American astronaut Kathy Sullivan made history in 1984 when she became the first American woman to complete a spacewalk. She was one of the first six women to be accepted as a NASA astronaut. Sally Ride, another of those six women, was the first American woman in space.

On June 7, 2020, Sullivan became the first woman to reach the deepest part of the Earth's oceans. She descended more than 35,800 feet (10,900 meters) in a two-person submersible to reach The Challenger Deep in the Mariana Trench. It's almost seven miles (11 km) below the ocean surface.[7]

To put that into perspective, Mt. Everest—the highest point on Earth—stands at 29,032 feet (8,848.86 meters) as of December 2020.[8] The Challenger Deep is deeper than Mt. Everest is tall.

In March of 2021, space tourist Richard Garriott became the second person in the world to visit both space and The Challenger Deep.[9]

By Comparison...

HOW MANY PEOPLE HAVE SUMMITED MT. EVEREST?

According to the Himalayan Database, 5,790 people have summitted Mt. Everest as of August 2020.[10]

WHO HAS BEEN TO THE BOTTOM OF THE OCEAN?

As of April 2021, only 21 people have been to The Challenger Deep[11], which is the deepest part of the Mariana Trench. James Cameron, the *Titanic* and *Avatar* director, was the first person to go to The Challenger Deep solo.[12]

HOW MANY PEOPLE HAVE BEEN TO THE EARTH'S POLES?

Now that there are tour operators that can fly people to both the South and North Poles, the number of people that have been to the Earth's Poles are estimated to be several thousand.[13]

THREE POLE CHALLENGE PARTICIPANTS
Just over 20 people have been to the North Pole, the South Pole and have summitted Mt. Everest.[14]

NOTES

1. "List of Space Travelers by Name," Wikipedia, last modified July 20, 2021, https://en.wikipedia.org/wiki/List_of_space_travelers_by_name.

2. Marianne Jennings, "Complete List of People Who Have Been to Space," KnowledgeNuggetBooks.com, October 24, 2021, https://knowledgenuggetbooks.com/complete-list-of-people-who-have-been-to-space/.

3. "Celebrating Women's History Month: Most Recent Female Astronauts," NASA, 2021, https://www.nasa.gov/mission_pages/station/research/news/whm-recent-female-astronauts.

4. "Visitors to the Station by Country," NASA, 2021, https://www.nasa.gov/feature/visitors-to-the-station-by-country/.

5. "Meet the Only NASA Astronaut to Climb Mount Everest," Business Insider, 2019, https://www.businessinsider.com/nasa-astronaut-climb-mount-everest-space-2019-4?r=US&IR=T.

6. Rajan Pokhrel, "After Flying in Space, Maurizio Cheli Becomes Second Astronaut to Climb Mt Everest," The Himalayan Times, 2018, https://thehimalayantimes.com/nepal/after-flying-in-space-maurizio-cheli-becomes-second-astronaut-to-climb-mt-everest/.

7. "Kathy Sullivan: The Woman Who's Made History in Sea and Space," BBC News, 2020, https://www.bbc.co.uk/news/world-us-canada-53008948.

8. "Mt Everest Grows by Nearly a Metre to New Height," BBC News, 2020, https://www.bbc.co.uk/news/world-asia-55218443.

9. "Astronaut-Explorer Richard Garriott Sets Records on Dive to Deepest Point on Earth," Collectspace.com, 2021, http://www.collectspace.com/news/news-030321a-astronaut-garriott-dive-challenger-deep.html.

10. "Everest 2021: Welcome to Everest 2021 Coverage," The Blog on Alanarnette.com, 2021, https://www.alanarnette.com/blog/2021/03/01/everest-2021-welcome-to-everest-2021-coverage/.

11. "List of People Who Descended to Challenger Deep," Wikipedia, 2021, https://en.wikipedia.org/wiki/List_of_people_who_descended_to_Challenger_Deep.

12. "James Cameron Completes Record-Breaking Mariana Trench Dive," National Geographic Adventure, 2012, https://www.nationalgeographic.com/news/2012/3/120325-james-cameron-mariana-trench-challenger-deepest-returns-science-sub/.

13. Joseph V. Micallef, "The North and South Poles Have Become the Ultimate Travel Bucket List: Here's How To Get There," Forbes, November 8, 2020, https://www.forbes.com/sites/joemicallef/2020/11/08/the-north-and-south-poles-have-become-the-ultimate-travel-bucket-list-heres-how-to-get-there/?sh=174b8a41e554.

14. Adventurestats.com, accessed May 11, 2021, http://www.adventurestats.com/statistics.shtml.

SPACE AGENCIES AROUND THE WORLD

Government Space Agencies

As of 2018, there are 72 different government space agencies around the world. Fourteen of these have launch capabilities and six of those have full launch capabilities.

THE BIG SIX

These space agencies all have full launch capabilities and Mission Control centers.

- **USA – NASA**, which stands for National Aeronautics and Space Administration
- **RUSSIA** – Roscosmos or RFSA, which stands for Russian Federal Space Agency
- **JAPAN – JAXA**, which stands for Japan Aerospace Exploration Agency
- **CHINA – CNSA**, which stands for China National Space Administration
- **EUROPEAN – ESA**, which stands for European Space Agency. These 17 countries are part of the ESA: Austria, Belgium, Denmark, Finland, France, Germany, Greece, Ireland, Italy, Luxembourg, the Netherlands, Norway, Portugal, Spain, Sweden, Switzerland and the United Kingdom.
- **INDIA - ISRO**, which stands for Indian Space Research Organization

DON'T FORGET CANADA

The Canadian Space Agency does not have full launch capabilities, but the country has played a big part in space exploration, sending several astronauts to space and developing several key pieces of equipment that have been critical to space exploration, like the Canadarm2.

Mission Controls Around the World

HOUSTON

NASA's Mission Control is in Houston, Texas at the Johnson Space Center. Launches are run at the Kennedy Space Center in Florida. But after launch, responsibility and communication moves to Houston.

TsUP

Russian Mission Control is referred to as TsUP (pronounced like "soup") and is located just outside Moscow, Russia. Daily communication with the ISS happens at TsUP—especially with the Russian portion of the ISS.

TKSC

The Japan Mission Control is at the Tsukuba Space Center in a city called Tsukuba, which is just outside of Tokyo. Anything that happens in the JAXA's Kibo ISS research lab or has anything to do with JAXA's satellites, is communicated through here.

COL-CC

ESA's mission control center is called Col-CC, which stands for The Columbus Control Centre and is near Munich, Germany. Support for European's Columbus research lab on the ISS is located here.

Facilities for Crewed Launches

There are several launch facilities—also called spaceports—around the world, and they fall into two different categories: launching satellites and launching humans. The following are the main spaceports for launching humans into space.

BAIKONUR COSMODROME IN KAZAKHSTAN

This is the world's first and largest operational spaceport. This is where the world's first satellite, Sputnik 1, was launched in 1957. It's also where the first human was sent to space in 1961 onboard Vostok 1.[1] Russia leases the Baikonur Cosmodrome from Kazakhstan. The current lease lasts until 2050. Russia launches both crewed and uncrewed spacecraft from the Baikonur Cosmodrome.

Spectacular Pyrotechnics

Other launch sites around the world store water under the launchpad, and spray it at the rockets during ignition to douse the flames and muffle the sound. Baikonur is way out in the desert of southern Kazakhstan, where water is scarce. So they skip that step, which makes for an impressive and fiery launch.

Originally a Missile Test Range

Having uninterrupted radio signals from the ground control stations to the object you launch is important—especially if it happens to be a missile. Baikonur, Kazakhstan, was originally a test range for the world's first intercontinental ballistic missile in 1955. This was the perfect location because it's out in the middle of nowhere, surrounded by flat plains.[2]

World's Largest Industrial Railway

All of Russia's spacecraft and ISS modules that are built in Moscow are transported about 1,300 miles (2,100 km) to the Baikonur launchpads by railway. The size of the Russian modules for the ISS were determined based on the diameter, height and width of the railway overpasses from Moscow, Russia to Baikonur, Kazakhstan.

U.S. KENNEDY SPACE CENTER

The Kennedy Space Center is America's spaceport and is named after President John F. Kennedy. It's located on Merritt Island in

Florida right next to Cape Canaveral, where NASA has launched Apollo, Gemini and Mercury crafts, the Space Shuttles and most recently SpaceX Falcon Dragon spacecraft.

Why Florida?

When they were chosen in the early 1960s, Cape Canaveral and Merritt Island were unpopulated areas with mostly undeveloped land. Much of Merritt Island is a restricted area and home to a wildlife sanctuary. It's not uncommon to come across bald eagles, wild boars, alligators, rattlesnakes, panthers and manatees. Another advantage is that it's close to the ocean, which meant objects launched would mainly be flying over the ocean and not where people live. Any debris or explosions would most likely fall over the ocean and not over people.[3]

Vertical Assembly

American spacecrafts are assembled vertically inside the 525-foot (160-meter)-tall Vehicle Assembly Building. It's one of the largest buildings in the world by volume. Once assembled, these spacecrafts are then transported to the launchpad in the same vertical position atop a massive crawler.

One of the main differences between Russian and American approaches to spacecrafts is that Russian space vehicles are assembled and transported horizontally and then stood up vertically once they reach the launchpad.

CHINA'S JIUQUAN SATELLITE LAUNCH CENTER

The Jiuquan Satellite Launch Center (JSLC) is the first and oldest of four spaceports for China. It's located in the Gobi Desert in the middle of Mongolia. It has a dedicated railway line that connects with their Technical Center and Launch Area. The Gobi Desert is dry almost all year long and has extended daylight hours. It has relatively low humidity and has an average temperature of 46 °F (8 degrees Celsius), which makes it well suited for launching rockets.[4]

CALIFORNIA SPACEPORT GOES HORIZONTAL

In a break in tradition, Virgin Galactic's suborbital SpaceShipOne spacecraft launches horizontally while attached to a carrier airplane. Its first launch happened in 2004 at California's Mojave Air and Space Port.

NEW MEXICO'S SPACEPORT AMERICA

It's known as the "World's First Purpose-Built Commercial Spaceport" which means that it was built for commercial uses, which aren't allowed at other airports or government-owned launch pads. It was built to support both horizontal and vertical launches. Virgin Galactic has moved all of its operations to Spaceport America. Besides Virgin Galactic, 10 other companies—including Google, Boeing and more—have used or will use Spaceport America's facilities for launch.

NOTES

1. Tim Peake, *Ask an Astronaut: My Guide to Life in Space* (Little, Brown and Company, 2017), 13.

2. Peake, *Ask an Astronaut*, 14.

3. "Why Did NASA Choose to Launch Rockets from Cape Canaveral, Florida?" *Space Coast Launches* (blog), accessed May 28, 2021, https://spacecoastlaunches.com/blog/why-did-nasa-choose-to-launch-rockets-from-cap-canaveral/.

4. "Jiuquan Satellite Launch Center," CGWIC, accessed May 28, 2021, http://www.cgwic.com/launchservices/launchsite/jslc.html.

SO YOU WANT TO BE AN ASTRONAUT?

Applying to Become an Astronaut

THE SELECTION PROCESS

Unlike applying to college, there is no fixed time frame when space agencies search for astronauts. It all depends on when new recruits are needed and which agency you are applying to.

Since 1959, NASA has selected astronauts 22 times as of April 2021. During the 1960s, new astronauts were needed about every 18 months. But from 1969 to 1978, no astronauts were selected. All that changed with the NASA Space Shuttle program. During the 1990s through 2000, astronauts were selected every two years for both Space Shuttle and International Space Station missions. After 2000, NASA has chosen astronauts about every four years. The most recent NASA astronaut applications were taken March 2-March 31, 2020.

The European Space Agency has selected astronauts three times: 1992, 1998 and 2009. Their most recent call for applications were at the end of March 2021.

The Japanese space agency has selected astronauts in 1985, 1992, 1998 and 2009.

The Canadian Space Agency has selected astronauts in 1983, 1992, 2008 and 2016. On average, applications open up every 8-10 years.

Russian cosmonauts are selected every 2.5 years on average.[1]

Basic Requirements

GENERAL KNOWLEDGE

Astronaut applicants are expected to have a basic level of knowledge in math, science, engineering and the English language. As part of the selection process, applicants are tested in these areas with tests that are designed to be stressful, with very few breaks in between tests. Candidates are expected to complete the test with speed and a high degree of accuracy. [2]

NASA EDUCATION REQUIREMENTS

NASA expects applicants to have at least a master's degree in engineering, biological science, physical science, computer science or mathematics and at least two years of professional experience. Those with pilot experience need to have at least 1,000 hours of flight time as pilot-in-command in a jet aircraft or have completed a test pilot program—or both! Those with their doctorate in medicine or osteopathic medicine are also eligible. [3]

ESA EDUCATION REQUIREMENTS

For the ESA or European Space Agency, astronaut applicants must have at least a master's degree in natural sciences, medicine, engineering, mathematics, or computer science, with at least three years of professional experience. A degree as an experimental test pilot and/or test engineer is also accepted. [4]

CSA EDUCATION REQUIREMENTS

For the CSA or Canadian Space Agency, astronaut applicants having at least a bachelor's degree in engineering, physics, chemistry, biology, geology, mathematics, or computer science is sufficient. They also accept candidates with a doctorate in medicine or dentistry. Applicants must also have at least three years of relevant professional experience and/or be licensed to practice medicine in Canada.[5]

MEDICAL REQUIREMENTS

Astronauts are living and working in space for longer and longer time periods. They need to have healthy bodies that won't require a trip to the doctor for 6 months to a year—or longer. This is part of why there are such strict medical requirements for astronauts.

The medical screening has more to do with choosing candidates who have a long-term prognosis for good health, which also means a low risk of developing medical problems while in space.[6]

EYESIGHT

Astronaut candidates need to have 20/20 visual acuity for both near and far vision in each eye, color perception and 3D vision. But wearing glasses or contact lenses does not disqualify someone from becoming an astronaut. The vision test requires that with glasses or contacts, their vision is corrected to see 20/20. Applicants who have had laser surgery to correct their vision will need to have their case looked at individually.[7]

HEARING

Astronauts have to hear well, especially since astronauts often work in environments with a lot of background noise. They must be able to communicate over the radio with ground control. Applicants have to have a hearing capacity of 25 decibels or better.[8] To put that into perspective, the sound of a mosquito or rustling leaves is 20 decibels. A whisper is normally at 30 decibels.[9]

MENTAL HEALTH

As space exploration continues to move forward, the need for people who can live with others for months and years at a time is critical. Currently, space programs are looking for people who will go to the Moon and eventually Mars. These will be years-long expeditions, and having people who can work with others and cope with stress in isolated and confined environments is critical. Space programs have looked at how people live and work in environments of Antarctica during the long winter as an example of what this type of environment might look like.

People who thrive in Antarctica—especially during the winter months—are thought to be mentally well-equipped to thrive in the isolation and confinement of space travel.[10]

YOU MUST BE THIS TALL TO RIDE

Due to logistics with spacesuits and spacecrafts, astronauts must measure between 5 feet 2 inches and 6 feet 3 inches tall (157 cm and 190.5 cm) and weigh between 110 and 209 lbs (50 and 95 kg).

NO CLAUSTROPHOBIA

Part of being an astronaut includes being comfortable in tight, confined spaces for long periods. Space capsules aren't the roomiest space vehicles, so one of the most basic requirements is not being claustrophobic.

Astronaut Scott Kelly shared how they tested candidates for claustrophobia when he was going through the training program. Each astronaut was fitted with a heart monitor, zipped into a thick rubber bag about the size of a curled-up adult, shut into a closet and then left without any idea of how long they'd be in there. For those who are claustrophobic, this is the stuff of nightmares. For those who aren't bothered, they may find it's a good time to enjoy a brief twenty-minute nap like Kelly did.[11]

PHYSICALLY FIT

Space agencies aren't looking for extreme levels of fitness or extreme athletes. They are looking for people who are fit and healthy. Astronauts who live and work in the International Space Station must work out at least two hours every day with cardio and resistance training.

NASA astronaut applicants will have to pass the NASA long-duration flight astronaut physical which involves knowing how to swim. They will be expected to pass a swimming test and military water survival training if they are selected, since some landings occur in water.

MUST BE CITIZEN

If you want to apply to NASA, you must be an American citizen or a dual citizen.[12] CSA requires applicants to be Canadian citizens or permanent residents of Canada and preference is given to Canadian citizens.[13]

ESA applicants need to be nationals from one of the 22 European member states and 3 associate member states. These include Austria, Belgium, the Czech Republic, Denmark, Estonia, Finland, France, Germany, Greece, Hungary, Ireland, Italy, Luxembourg, the Netherlands, Norway, Poland, Portugal, Romania, Spain, Sweden, Switzerland and the United Kingdom plus Slovenia, Latvia and as of May 20, 2021, Lithuania.[14]

Unique Canadian Space Agency Astronaut Selection Process

SWIMMING IS ESSENTIAL

Canadian astronaut applicants are given a swim test. To pass, they must swim at least 820 feet (250 meters) in 10 minutes, tread water for at least 10 minutes and swim at least 49 feet (15 meters) underwater.[15]

EMERGENCY SIMULATION TESTS

Canadian applicants who reach a certain level are put through emergency simulations in order to test how resilient they are, in addition to how well they think and work under pressure, make decisions, work as a team and solve problems. These tests include fighting fires, diving to the bottom of pools to retrieve bricks and working as a team to plug leaks in a room that is being flooded with cold water from the Atlantic Ocean.[16]

One of the emergency simulation tests, called "The Dunker Test," involves strapping astronaut hopefuls into a mock helicopter cockpit, giving last-minute escape instructions and then letting the cockpit crash into a swimming pool. They have only seconds before the cockpit fills with water and capsizes. This tests their ability to stay calm in the face of panic, follow instructions and escape without becoming disoriented.[17]

PUBLIC SPEAKING & SOCIAL MEDIA

Astronauts become role models and ambassadors for their

space agency. They are asked to speak to kids in schools and at business events. Communication skills are critical. As part of the selection process, candidates give presentations, give interviews with the media, film video messages and produce content for social media.[18]

REALITY TV OR ASTRONAUT SELECTION?

In the most recent astronaut selection campaign, the Canadian Space Station shared a series of videos with the public at each stage of the selection process. The videos, called "Astronauts Wanted," gave viewers the opportunity to watch as candidates were put through a number of tests and evaluations.[19]

Unique Japanese Astronaut Selection Process

ISOLATION CHAMBERS

For two openings in the Japanese astronaut corps, 10 finalists spend one week together in an isolation chamber. Isolation chambers are a way to test group skills that can't be determined in one-on-one interviews. These skills include teamwork, leadership and conflict management.[20]

THE THOUSAND CRANES TEST

To test how astronaut finalists handle impatience, stress and tedious work, the JAXA program came up with the Thousand Cranes Test. Each finalist must fold 1,000 origami cranes and thread them onto a single long thread over the course of a few days.

This task becomes a chronological record of each finalist's work. Japanese psychologists analyze how the last cranes compare to the first ones the finalists made. Do the candidates get sloppy with their cranes as the deadline approaches and the pressure increases? In the words of JAXA psychologist, Natsuhiko Inoue, "Deterioration of accuracy shows impatience under stress."

Much of what astronauts do on the International Space Station involves assembling, repairing and maintaining the spacecraft. It can be tedious work under pressure (figuratively and literally) and with a clear deadline. These tasks are often done while in a pressurized suit and with a limited supply of oxygen. Having astronauts who can handle stress, impatience and tedious work all while maintaining accuracy and the quality of their work is vital.

 DID YOU KNOW?
According to Japanese tradition, a person who folds one thousand cranes will be granted health and longevity. This gift of health and longevity can be passed onto others. Cranes are often seen strung on strings and given to hospital patients.[21]

Disqualifications

Going into space, living in close quarters with others, working under stress and being an ambassador for a space agency comes with high standards. Many conditions can disqualify a person from the different space agencies, including some you may not have considered. In some cases, if a person is disqualified they can apply in the future if they are able to resolve the reason they were originally disqualified.

SNORING
According to JAXA, if it's loud enough and wakes people up, this could mean disqualification.[22]

BAD BREATH
If you've ever had to share a tight space with someone with bad breath, you can see why this would cause elimination. The Chinese space agency will disqualify an applicant with bad breath because "the bad smell would affect their fellow colleagues in a narrow space."[23]

COLORBLINDNESS

Currently, the Canadian Space Agency will not accept applicants who are colorblind.[24] In the past, NASA did not accept those who were colorblind. But that agency has made some exceptions for certain positions during the space program. Astronaut Roger Crouch, a payload specialist during the space shuttle era, is colorblind.[25]

NOT ABLE TO READ, WRITE AND SPEAK ENGLISH

The international languages on the ISS are English and Russian. Knowing English is a requirement, but knowing Russian isn't. Astronaut candidates will learn Russian during their training.

DRUGS, ALCOHOL AND TOBACCO DEPENDENCY

Applicants who have an addiction to drugs, alcohol or tobacco won't make it far. There is a no-alcohol policy on the ISS, and tobacco isn't part of the space-safe items onboard the ISS.[26]

HIGH BLOOD PRESSURE

Applicants must have blood pressure that is lower than 140/90 while sitting. Generally, blood pressure readings between 90/60 and 120/80 are considered normal.[27]

VACCINATIONS

Astronauts are required to have and get vaccinations for communicable diseases. Those who aren't vaccinated and aren't willing to get vaccinated won't be selected.[28]

AGE LIMITS

NASA doesn't currently have an age limit for their applicants, but the ESA has an age limit of 50.[29]

DRIVER'S LICENSE

Astronauts need to be able to get themselves around to a variety of places. Having a driver's license and being able to drive is now required. If the applicant doesn't have a driver's license, they do need to be willing and able to get one.[30]

FEAR OF WATER
Swimming is a must. Astronauts spend hundreds of hours underwater training for spacewalks. People who have a fear of being in the water won't make it very far.

 DID YOU KNOW?
IF AT FIRST YOU DON'T SUCCEED, TRY AGAIN
Several professional astronauts were not accepted as astronaut candidates the first time they applied. For some, it took many times. Astronaut Mike Massimino was rejected 3 times before he was accepted.[31] It took Peggy Whitson, the woman who has spent the most time in space, applying 5 times over 10 years before she was accepted.[32] Clay Anderson applied to become an astronaut 15 times before even getting an interview with NASA. It then took another two years before he was selected as an astronaut candidate.[33]

Dangerous Job

ODDS OF DYING ON THE JOB
As of July 2021, 574 people, both men and women, have been to space. Of the 574 people, 30 have died while flying to space, returning to Earth or preparing to fly to space. Examples include the *Challenger* and *Columbia* disasters and the tragic fire that killed all of the astronauts during a test run on the launchpad for Apollo 1. Statistically, that means that the odds of an astronaut dying on the job is about 5%, or 1 in 20.[34] However, that number is inflated because in reality, several people have flown multiple times which reduces the fatality rate.[35]

FIRE

Fire is a real concern in space, since it's an oxygen-rich environment. Oxygen is highly flammable. Fires have happened several times in the history of space flight. The fire that broke out on Apollo 1 killed everyone onboard, changing NASA rules and protocols to reduce the risk of that ever happening again. In 1997, a fast-moving fire onboard space station Mir blinded and choked the crew within seconds. Luckily, the team was able to react quickly and put the fire out, and no one died.

Cosmonaut Sasha Kaleri was cutting his hair when the fire started on Mir. Some cosmonauts, especially the older ones, are superstitious. They refuse to cut their hair while in space.[36]

ELECTROCUTION

Onboard the International Space Station there is a risk of electrocution, especially on the side managed by the U.S. The U.S. side uses 120-volt power, which is much more dangerous than the 28 volts used on the Russian side. When ISS crews clean or have to work in certain areas, they are careful to turn the power off in those areas or on a particular part to prevent a shortage or electrocution.[37]

ASTRONAUT INSURANCE & AUTOGRAPHS

Being an astronaut is a dangerous job, but was especially so in the early days of space flight. Getting life insurance as an astronaut was almost impossible, unless a person was willing to pay astronomical prices. Astronauts figured out they had something most everyone wanted and would find valuable: their autograph.

Apollo 11 astronauts Neil Armstrong, Buzz Aldrin and Michael Collins signed hundreds of envelopes and gave them to a friend before they flew to the Moon. The friend was instructed to mail the envelopes to the crew's families on special dates, especially on launch day and the day they landed/walked on the Moon. These envelopes are known as "covers" and they were to be sold to the highest bidder to provide money for the crews' family in case they didn't return or returned but could no longer work. The cost of an Apollo 11 autograph can run up to $30,000 in today's money.

The practice of using "covers" as astronaut insurance may have started as early as NASA's Mercury program in the early 1960s.[38]

The Backup Crew

NASA selected backup crews for every launch in the Mercury, Gemini and Apollo programs. Every launch had a prime crew and a backup crew. The backup crew was there in case one or all of the prime crew wasn't able to go through with a launch. NASA discontinued the practice of having a backup crew for the entire prime crew after STS-3 in the Shuttle program.[39] There are still backups for individual crew members at times, but not for the entire crew.

It's also common for cosmonauts to have backup crews in case anything happens to one of their prime crew members.

SELECTED AT THE SAME TIME

Usually, a prime crew and backup crew for a particular mission were selected at the same time so they could start the training process together. This allowed everyone to be on the same page from the beginning.[40] Like an understudy in the theater, the backup crew was available to step in if needed.

THEY NEVER TRAVEL TOGETHER

Just like the president and vice president of the United States, the prime and backup astronaut crews always traveled separately.

A CASE OF MEASLES

Just 72 hours before Apollo 13 was to launch, command module pilot Ken Mattingly was grounded and replaced by Jack Swigert, who was on the backup crew. Ken had been exposed to measles and NASA didn't want to risk him getting the measles during the trip. Ken never did get the measles, but he did get to fly to the Moon as the command module pilot on Apollo 16.

NOTES

1. Robert Frost, reply to "How often do space agencies, (NASA, ESA & ROSCOSMOS) recruit astronauts/cosmonauts?" Quora, accessed May 11, 2021, https://www.quora.com/How-often-do-space-agencies-NASA-ESA-ROSCOSMOS-recruit-astronauts-cosmonauts.

2. Tim Peake, *Ask an Astronaut: My Guide to Life in Space* (Little, Brown and Company, 2017), 55.

3. Flint Wild, "Astronaut Requirements," NASA, last updated January 29, 2021, https://www.nasa.gov/audience/forstudents/postsecondary/features/F_Astronaut_Requirements.html.

4. "Astronaut Selection 2021-22 FAQs," European Space Agency, accessed May 11, 2021, https://www.esa.int/About_Us/Careers_at_ESA/ESA_Astronaut_Selection/Astronaut_selection_2021-22_FAQs.

5. "Requirements and Conditions of Employment for Astronauts," Canadian Space Agency, June 26, 2008, https://www.asc-csa.gc.ca/eng/astronauts/how-to-become-an-astronaut/requirements-and-conditions.asp.

6. Peake, *Ask an Astronaut*, 57.

7. "Astronaut Selection 2021-22 FAQs."

8. "Astronaut Selection 2021-22 FAQs."

9. "Decibels Explained," TLC Electrical, accessed September 1, 2021, https://www.tlc-direct.co.uk/Technical/Sounds/Decibles.htm.

10. Mary Roach, *Packing for Mars: The Curious Science of Life in the Void* (New York: WW Norton, 2011), 31.

11. Scott Kelly, *Endurance: A Year in Space, a Lifetime of Discovery* (New York: Alfred A. Knopf, 2017), 203.

12. Wild, "Astronaut Requirements."

13. "Requirements and Conditions of Employment for Astronauts."

14. "Astronaut Selection 2021-22 FAQs."

15. "Frequently Asked Questions," Canadian Space Agency, last modified September 1, 2020, https://www.asc-csa.gc.ca/eng/astronauts/faq.asp.

16. "Meet Canada's Two New Astronauts," Canadian Space Agency, last modified July 1, 2017, https://www.asc-csa.gc.ca/eng/astronauts/how-to-become-an-astronaut/2017-recruitment-campaign.asp.

17. "2009 Astronaut Recruitment Campaign," Canadian Space Agency, last modified June 17, 2016, https://www.asc-csa.gc.ca/eng/astronauts/how-to-become-an-astronaut/2009-recruitment-campaign.asp.

18. "Meet Canada's Two New Astronauts."

19. "Astronauts Wanted – Episode 1: Launch of Canada's Fourth Astronaut Recruitment Campaign," Canadian Space Agency, June 17, 2016, video, https://www.asc-csa.gc.ca/eng/search/video/watch.asp?v=1_pfuugodj.

20. Roach, *Packing for Mars*, 24.

21. Roach, *Packing for Mars*, 25-26.

22. Roach, *Packing for Mars*, 33.

23. Roach, *Packing for Mars*, 34.

24. "Requirements and Conditions of Employment for Astronauts."

25. JoCasta Green and David Hitt, "The Power of Persistence," NASA Educational Technology Services, April 4, 2016, https://www.nasa.gov/audience/forstudents/k-4/home/F_The_Power_of_Persistence_K-4.html.

26. J. Coise and J. Grandsire, European Astronaut Selection (Germany: European Space Agency, 2008), pdf, https://www.esa.int/esapub/br/br271/br271.pdf.

27. "What Is Blood Pressure?" NHS, accessed September 1, 2021, https://www.nhs.uk/common-health-questions/lifestyle/what-is-blood-pressure.

28 "Requirements and Conditions of Employment for Astronauts."

29 "Astronaut Selection 2021-22 FAQs."

30 "Astronaut Selection 2021-22 FAQs."

31 Zameena Mejia, "The No. 1 Lesson on Persistence from an Astronaut Who Got Rejected by Nasa 3 Times before Getting the Job," CNBC, April 13, 2018, https://www.cnbc.com/2018/04/13/nasa-rejected-this-former-astronaut-3-times-before-giving-him-the-job.html.

32 Peggy Whitson, "How 10 Years of Rejection by NASA Taught Peggy Whitson to Be a Better Astronaut," interview by Kate Snow, NBC Nightly News, 2019, https://www.youtube.com/watch?v=naWbcPFejAw

33 Brian Dunbar and Amiko Nevills, "Clay Anderson: The Little Engine That Flew," NASA, November 23, 2007, https://www.nasa.gov/astronauts/anderson_profile.html.

34 Corey S. Powell, "Why Astronaut Chris Hadfield Isn't Afraid of Death," NBCNews.com (NBC Universal News Group, July 4, 2018), https://www.nbcnews.com/mach/science/why-astronaut-chris-hadfield-isn-t-afraid-death-ncna860511.

35 "Weighing the Risks of Human Spaceflight," The Space Review, accessed May 12, 2021. https://www.thespacereview.com/article/36/2.

36 Kelly, Endurance, 101.

37 Kelly, Endurance, 125.

38 Merryl Azriel, "Neil Armstrong's Insurance Policy," Space Safety Magazine, August 31, 2012, http://www.spacesafetymagazine.com/spaceflight/life-in-orbit/neil-armstrongs-insurance-policy/.

39 "Astronaut Killed in Air Crash May Have Broken NASA Rule," Los Angeles Times, June 19, 1989, https://www.latimes.com/archives/la-xpm-1989-06-19-mn-1869-story.html.

40 "Assigning an Astronaut to a Mission," European Space Agency, accessed May 12, 2021, https://www.esa.int/Science_Exploration/Human_and_Robotic_Exploration/Astronauts/Assigning_an_astronaut_to_a_mission.

ASTRONAUT CANDIDATES & BASIC TRAINING

THE NEW RECRUITS

AsCans (pronounced ass-cans) is the nickname given to astronaut applicants who have been selected and are now officially Astronaut Candidates.

BASIC TRAINING PHASE

Basic training starts as soon as applicants become astronaut candidates. They are assigned to the Astronaut Office at the Johnson Space Center (JSC) in Houston, Texas. Together, basic training and the evaluation period lasts approximately two years.[1]

PRE-ASSIGNMENT PHASE

After basic training, most astronauts have to wait a few years before they are assigned to their first space mission. While they wait, their job is to support the human spaceflight program and continue to train and enhance their own knowledge, operational skills and professional development.[2] They often work as CAPCOM at mission control (the person who is responsible for talking to the astronauts in space), help train future astronauts and work to develop the software and technology astronauts will use in future missions.

ASSIGNED TRAINING FLOW PHASE

This final phase of training happens once an astronaut has been assigned to a mission. This usually lasts about 2 ½ years, or until launch.

PROMOTION TO ASTRONAUT

Astronauts officially become astronauts once they fly into space. Until then, they're astronaut candidates.[3]

A LONG WAIT

Most astronauts will have been waiting and training a minimum of three to four years before their first flight into space.[4]

14 YEARS OF WAITING

Swedish astronaut Christer Fuglesang waited 14 years for his first mission to space in 2006.[5]

18 YEARS OF WAITING

NASA Astronaut Bruce McCandless—who became the first astronaut to free-fly with a jet-propelled backpack—waited 18 years before his first spaceflight in 1984. He was selected for astronaut training in 1966 during the Apollo era. During his time waiting for his flight assignment, Bruce worked to develop the Manned Maneuvering Unit (MMU), the same jet-propelled backpack he was the first to fly in space.[6]

THE LONGEST WAIT WAS 19 YEARS

Selected in 1966, NASA Astronaut Don Lind waited for 19 years before he made it into space in 1985. Lind and Bruce McCandless were both selected as part of NASA's Astronaut Group 5 and were treated as scientist-astronauts. They had to wait a long time before they flew into space perhaps due to their lack of test pilot experience. Lind's one and only space mission was onboard STS-51-B on *Challenger*. Lind holds the record for having the longest wait between astronaut selection and space flight for an active astronaut.[7]

EARLY PROMOTION

There are those who don't have to wait long at all. In 2011, Italian astronaut Luca Parmitano was assigned his first mission while he was still in basic training![8]

Astronaut Basic Training

DIFFERENT TRAINING FOR DIFFERENT SPACE AGENCIES

Basic training isn't the same for all space agencies, but there is an agreed training criteria that all astronaut candidates learn.[9]

SWIM TEST

In the first month of basic training, all astronaut candidates are required to pass a swimming test. This involves swimming 3 lengths of a 25-meter pool without stopping, then swimming another 3 lengths of the pool in a flight suit and tennis shoes with no time limit. They must also tread water continuously for 10 minutes wearing a flight suit.[10]

SCUBA CERTIFICATION

In order to train and prepare for a spacewalk, hundreds of hours are spent in the NBL pool both as a diver and in a spacesuit. If they aren't already certified, candidates get scuba certified.

PHYSICALLY FIT

Astronauts are expected to have at least 4 hours of structured physical training each week in addition to any personal sports. Most people do more than this though.

ACADEMIC STUDIES

Astronaut candidates come from a variety of backgrounds. They need to transition from being a specialist in whatever area they studied to a generalist, who knows about many things. Candidates have a series of short, intensive courses in subjects like biology, geology, meteorology, information networks and rocket science. They have to pass a final exam for each subject. This helps the candidates have the same foundation of knowledge, which helps them communicate with each other and with the scientists and technicians they'll be working with.[11]

ROCKET SCIENCE

As common sense as it may seem, astronauts have to understand how rockets work. This involves understanding what it takes to get a rocket into space and into orbit, how fast it needs to go to escape Earth's gravity, how it's designed, what the controls mean and even how the weather and atmosphere are involved.

ORBITAL MECHANICS

Astronauts need to learn how things orbit the earth and then learn how to actually do it as an astronaut when flying into space.

OBSERVER OF THE WORLD

Astronauts have a unique view of the world when they get to space and have the opportunity to make important observations. Knowing about geology, geography, volcanoes, land patterns, hurricanes, clouds or similar natural phenomena is important. This allows them to recognize what they see and report back their findings.

POST-LANDING SURVIVAL TRAINING

In case things go wrong in orbit and astronauts have to return to Earth, which is known as a deorbit, they have to know how to survive no matter where they land. The landing spot could be in the arctic, a desert, an ocean or anywhere in between. Astronauts train for these different scenarios so that they can stay alive until help arrives.

WATER SURVIVAL TRAINING

Water covers about 70% of Earth's surface. So chances are that an emergency deorbit landing will take place in water. Astronauts train getting into their water gear and into their emergency dinghy. They learn key skills like using shark repellant, communication with a radio and GPS and getting into the harness that a helicopter would drop down during a water rescue.

TEAM-BUILDING AND LEADERSHIP SKILLS

The survival training astronauts receive isn't just how to survive in emergency situations, but also how to lead and work effectively with their crew in those situations. Crews learn how they react to stress and then how to cope with it together. Survival training allows them to learn some of those critical lessons in a somewhat safe and controlled environment.

EXPLORING CAVES

ESA provides a caving course for their astronaut candidates. Candidates spend many days and nights living as a group of six in a vast cave network. The caves are technically challenging. It takes several hours of vertical ascents and descents using ropes and climbing equipment to arrive at base camp. Candidates explore deeper into the cave and collect microbiological samples and conduct scientific research, learn about cave photography and complete other tasks that involve teamwork and communication.[12]

NEEMO

NEEMO stands for NASA Extreme Environment Mission Operations. NASA sends groups of astronauts, engineers and scientists to live in Aquarius—an underwater laboratory off the coast of Florida—for up to three weeks at a time to prepare for future space exploration.[13]

Aquarius is the world's only undersea research station. It's used for experiments, technology tests and simulating several types of space operations. Astronauts can experiment with larger pieces of equipment like deepwater submersible vehicles.[14]

LANGUAGE TRAINING

The official languages onboard the International Space Station are Russian and English. Since 2011 and until 2020, the only way astronauts and cosmonauts got to the ISS was via the Russian Soyuz spacecraft. The instruments, control panels and flight documentation are all in Russian. Plus, the crew communicates with Moscow's Mission Control Center in Russian. The only English spoken in the Soyuz is the informal conversation between the crew.[15]

As soon as astronauts start basic training, they start learning Russian. The European Space Agency include a month-long immersion course living with a Russian family in St. Petersburg. Astronauts who are flying into space must pass the ACTFL (American Council on the Teaching of Foreign Language) oral-proficiency interview in Russian at the 'Intermediate High' level.

BEYOND LANGUAGE TRAINING

Learning the language is important, but so is understanding the Russian culture. The astronauts spend a lot of time in Star City in Russia where they learn all about the Russian segment of the ISS. While they are there, they also spend time with their cosmonaut colleagues at cultural and social events.[16]

MEDICAL TRAINING

While they are not given the full training that a nurse or professional emergency medical technician (EMT) would have, astronauts are trained to handle the basic first aid and other life-saving skills. They learn how to reinflate a collapsed lung or how to draw blood—even their own! They learn how to treat severe burns and to how to remove a metal splinter from an eyeball. They know how to perform various medical scans, like taking ultrasound measurements on the brain, the heart and eyeballs, all of which they perform on themselves monthly while in space.[17]

They are trained to handle medical emergencies and to be able to communicate with the flight surgeon in mission control. They have to know how to be the flight surgeon's eyes, ears and hands for any patient while in space.[18]

PHOTOGRAPHY TRAINING

Astronauts are observers of the world. They take a lot of photos and video. Some photos are required as part of experiments, some are engineering data that is given to flight controllers on the ground when something breaks, and some photos are taken because the astronaut feels like the shot was worth capturing. Every astronaut gets extensive photography and videography training. They learn how to set up the shots, keeping in mind factors such as exposure, focus, framing, sound and lighting.[19]

FILMMAKING & VIDEOGRAPHY TRAINING

Since the 1980s, more than 150 astronauts have been trained to be filmmakers by two of the best in the business: Toni Myers and James

Neihouse. Meyers was an award-winning documentary director and Neihouse is an amazing director of photography. If you've ever seen a space IMAX movie, you've seen their work.

SOUND LESSONS

Ben Burtt was the sound designer for many of the well-known *Star Wars* sounds. He has also worked with astronauts to make sure they capture the sounds of the space station. He worked with astronaut Terry Virts, who played a huge part by capturing video for the IMAX film, *A Beautiful Planet*, while in space. Together, Burtt and Virts captured space-station sounds, such as the fans whirring in the background, the metal workout equipment banging together like wind chimes and the sound of air rushing out into the vacuum of space through a hatch.[20]

IT'S NOT WHAT YOU SEE, IT'S WHAT YOU HELP OTHERS SEE

Astronauts and cosmonauts have had a unique vantage point. Over the years, they have captured over 3 million photos.[21] Astronaut Terry Virts holds the record for the most photographs taken while in space. During his 199 days onboard the ISS, he took over 319,000 photos!

BASIC ELECTRONICS REPAIR

Unfortunately, there is no engineer like *Star Trek*'s Scotty aboard the ISS. All of the astronauts and cosmonauts on the space station learn to do basic repairs like soldering and anything they'd need to do in an experiment repair shop. It's the crew that has to maintain the station and fix anything that goes wrong. The astronauts will have help from mission control, but it's their hands that do the fixing.[22]

NOTES

1. Brian Dunbar and Shelly Canright, "Astronauts in Training," NASA, April 9, 2009, https://www.nasa.gov/audience/forstudents/5-8/features/F_Astronauts_in_Training.html.

2. Tim Peake, *Ask an Astronaut: My Guide to Life in Space* (Little, Brown and Company, 2017), 62.

3. Scott Kelly, *Endurance: A Year in Space, a Lifetime of Discovery* (New York: Alfred A. Knopf, 2017), 207.

4. Peake, *Ask an Astronaut*, 61.

5. Peake, *Ask an Astronaut*, 62.

6. Peake, *Ask an Astronaut*,177.

7.   Don L. Lind, "Oral History," interview by Rebecca Wright, NASA Johnson Space Center Oral History Project, May 27, 2005, https://historycollection.jsc.nasa.gov/JSCHistoryPortal/history/oral_histories/LindDL/LindDL_5-27-05.htm.

8. Peake, *Ask an Astronaut*, 62.

9. Peake, *Ask an Astronaut*, 61.

10. "Astronaut Selection and Training," NASA, November 2011, https://www.nasa.gov/centers/johnson/pdf/606877main_FS-2011-11-057-JSC-astro_trng.pdf.

11. Samantha Cristoforetti, *Diary of an Apprentice Astronaut*, trans. Jill Foulston (Penguin, 2020), 27, Kindle.

12. Peake, *Ask an Astronaut*, 60.

13. Cristoforetti, *Diary of an Apprentice Astronaut*, 384.

14. Peake, *Ask an Astronaut*, 69.

15. Peake, *Ask an Astronaut*, 63.

16. Peake, *Ask an Astronaut*, 63.

17. Terry Virts, *How to Astronaut: An Insider's Guide to Leaving Planet Earth* (New York: Workman, 2020), 42.

18. Virts, *How to Astronaut*, 39.

19. Virts, *How to Astronaut*, 110.

20. Virts, *How to Astronaut*, 110.

21. Robert Frost, reply to "Why do we actually keep people in the ISS? Have we not exhausted all the possible worthwhile experiments that require humans in space?" Quora, accessed May 12, 2021, https://qr.ae/pNJyYb.

22. Robert Frost, reply to "Is there a 'Scotty' on the ISS, someone that is there to maintain the station and fix anything that goes wrong?" Quora, accessed May 12, 2021. https://qr.ae/pGnx2Y.

PROFESSIONAL ASTRONAUTS

What Astronauts Do When Not Assigned to a Space Mission

PRE-ASSIGNMENT PHASE

After basic training, most astronauts have to wait for a while—sometimes years—before they are assigned to their first space mission. While they wait, their job is to support the human spaceflight program and continue to learn and train as much as they can so they're ready when it's their time to go to space.[1]

MISSION CONTROL

Astronauts often spend time in Mission Control before and after their missions to space. They are often assigned as CAPCOM to be the designated voice to talk to the astronauts that are in space. During the Apollo and Shuttle missions, astronauts were the only ones assigned as CAPCOM. These days, astronauts perform CAPCOM duties during events like docking, and when astronauts leave the spacecraft on an EVA.

Charlie Duke, who walked on the Moon as part of the Apollo 16 crew, served as CAPCOM for the Apollo 11 mission—the mission to successfully land on the Moon. When Neil Armstrong radioed "Houston, Tranquility Base here. The Eagle has landed," Duke responded: "Roger. Tranquility. We copy you on the ground. You got a bunch of guys about to turn blue. We're breathing again. Thanks a lot."[2]

Sally Ride, the first American woman in space, worked as CAPCOM for the second and third shuttle flights. Those happened years before she flew to space in 1983.[3]

Astronaut Bruce McCandless II served as CAPCOM for several missions, including the Apollo 11 EVA, Skylab 3 and Skylab 4.

CAPCOM, EUROCOM, J-COM

This person is the voice link between the teams in space and the ground teams. Responsible for consolidating all of the information coming from a variety of sources on the ground and communicating it in a clear and concise way to the ISS crew.

CAPCOM comes from the early days of space flight, when the astronauts flew in capsules and there was a designated person to communicate with the capsule. CAPCOM became the shortened version of capsule communicator.

NASA's CAPCOM is often referred to as Houston/CAPCOM. EUROCOM is the European Space Agency's version of CAPCOM. J-COM is the Japanese Space Agency's version. The Russians refer to theirs as TsUP (pronounced like "soup") CAPCOM.[4] TsUP is an acronym for the name of Russia's Mission Control center just on the outskirts of Moscow, Russia.

HELP WITH DEVELOPMENT

Several astronauts are also involved in developing procedures, tools, systems and other things used in space. For example, Sally Ride and Judith Resnik helped develop the mechanical arm on the space shuttle that Ride then operated when she flew to space.[5] Bruce McCandless II was key in developing the jet pack called the Manned Maneuvering Unit (MMU). The MMU allowed astronauts to spacewalk untethered from the space shuttle. McCandless was the first to use it when he went to space for the first time 18 years after he was selected as an astronaut.

CHOOSING WAKE-UP MUSIC

One of the duties of CAPCOM has been to choose the wake-up music for those in space.

PUBLIC FACE OF THE AGENCY

Astronauts and Cosmonauts become the public face of their respective space agencies and are expected to talk about everything those space agencies do. NASA astronauts go on field trips to different NASA centers like the Ames Research Center in California, Glenn Research Center in Ohio, Goddard Space Flight Center in Maryland, Michoud Assembly Facility in Louisiana, Marshall Space Flight Center in Alabama, NASA headquarters in D.C. and the Kennedy Space Center in Florida. They learn what happens at each of those sites and how all of the NASA projects work together so they can then share it with others.[6]

Practice Makes Perfect

LEARNING TO SPACEWALK

For every hour an astronaut spends in space spacewalking, which is anytime an astronaut leaves the safety of the spacecraft while in space, the astronaut spends about 100 hours in a pool practicing everything they will do when they get to space. They train to maneuver in a spacesuit, operate the tools they'll use and follow the path they'll follow during their scheduled EVA. Spacewalking is one of the most dangerous jobs for an astronaut to do, but one of the most rewarding.

LEARNING TO USE THE ROBOTIC ARM

The ISS has a robotic arm that is very useful, but someone has to control it. Astronauts are trained to use the robotic arm, which is known as Canadarm 2, since it was built in Canada. The first Canadarm was used on the Space Shuttle.

RUN SIMULATIONS FOR EVERYTHING

To prepare and train for the experience of space, launch, life on the ISS, re-orbit and everything in between, astronauts simulate as much as they can. Astronaut Chris Hadfield has said, "The prep is what matters. That's what we do for a living. We don't fly in space for a living. We have meetings, plan, prepare, train. I've been an astronaut for six years and I've been in space for eight days."[7]

SERIOUSLY. EVERYTHING.

Apollo 13 astronaut Jim Lovell "simmed" on the ground everything he did in space, including not taking a bath for two weeks.[8]

Astronauts are given adult diapers to take home. They are told to put them on, and practice going to the bathroom while lying down in their bathtub.

Using the space toilets is also simulated. There's a replica where astronauts practice and perfect their aim.

SIMULATING WORST-CASE SCENARIOS

During a death simulation, also called contingency sims, the entire team works together to examine, react and work through possible scenarios. The team includes crewmates, Mission Control, public relations and management. Although it may seem morbid, these sims can also help the families of the crew better prepare for a dreaded situation. During one contingency sim, Astronaut Chris Hadfield brought his wife Helene along, so that she could accurately see how everyone would react and what her role would need to be. As a result, she modified her own plans for the duration of Chris's spaceflights.[9]

ROSCOSMOS SIMULATORS

In order to train cosmonauts and astronauts in an environment as real as possible, the Russians built an entire space station replica inside a vacuum chamber. When they simulate an air leak, cosmonauts and astronauts don't just see the pressure dropping, they can feel it dropping as their ears pop to adjust. This adds a whole level of urgency and realness to the training that studying the theory will never convey.[10]

USING VIRTUAL REALITY TO TRAIN

Virtual reality has made it possible to simulate things we wouldn't be able to otherwise, like being thrown off the space station and learning how to use the emergency jetpack to make it back. Astronauts wear virtual reality headsets with gloves that connect to the system that simulates what it looks and feels like to be working outside the space station with Earth spinning below them. Virtual reality training has already helped train astronauts for what to expect on their missions and will continue to do so as we get ready to send people to the Moon and Mars.[11]

Pay By Space Agency

The pay for becoming an astronaut or cosmonaut varies depending upon the space agency's country, the person's years of experience, military background and level of service. It's hard to compare between each space agency because tax structures, health insurance models and the standards of living are different in each location.

NASA SALARIES

NASA astronauts normally fall within two different categories: civilian and military.

Civilian astronauts are government employees and their salaries are based on the Federal Government's General Schedule pay scales for grades GS-12[12] up through GS-14.[13] The grade then depends on the astronaut's academic achievements and work experience. Currently a new astronaut with a GS-12 grade and very little experience starts at $66,829[14] per year. But an astronaut with years of experience at GS-14 can earn up to $161,141 per year.[15]

Astronauts with a military background, and active service members, bring with them most or all of their leave, bonuses, housing allowances and other military-grade benefits. Military officers' salaries are based on their years of service, typically Officer pay grades O-4 to O-6.[16] That can range from the low end of O-4 at $5,135[17] to the high end of O-6 at $12,638 a month.[18]

PAY IN CANADA

Canadian astronauts have three salary levels: Entry level, Qualified level and Senior level. The Entry level includes astronaut training. The Qualified level is once astronauts graduate from NASA astronaut basic training or the equivalent, and are fully qualified and awaiting their space mission assignment. The top level is the Senior level and is reached when the astronaut has successfully completed a space mission. The range is from $91,300 to $178,400 CAD.[19]

ESA EARNINGS

European astronauts use the A grade pay salary, usually A2 to A4 which ranges from 5,554.93 euros per month to 9,907.48 euros per month depending on experience and where you live.[20] ESA salaries are exempt from national income tax in ESA member states.[21]

CASH FOR COSMONAUTS

Russian cosmonauts are paid a base salary and then get paid bonuses for each day they fly in space. According to sources who work for NASA and are familiar with the different space agencies, a new cosmonaut pay can range from $1,830 to $2,640 per month. They then are paid to live and work on the International Space Station. A 6-month flight contract could be somewhere between $130,000 to $150,000.[22] Cosmonauts are docked pay for complaining, so they don't complain.[23]

SALARIES IN JAPAN

While the Japan Space Agency doesn't publish their salaries online, salary estimate websites who collect the information from employers and anonymous employees suggest that Japanese astronauts are paid based on their level of experience.

A new astronaut with 1 to 3 years of experience can earn an average base salary of ¥4,255,378 per year. On the other end of the scale, a senior level astronaut with eight or more years of experience can earn an average annual base salary of ¥7,246,130.[24]

This is equivalent to $39,146 USD to $66,659 USD per year. On top of their base salary, Japanese astronauts can earn an average bonus of ¥135,115 or $1,242 USD.

NOTES

1. Tim Peake, *Ask an Astronaut: My Guide to Life in Space* (Little, Brown and Company, 2017), 61.

2. Olivia B. Waxman, "Inside Mission Control during the Apollo 11 Moon Landing," Time, July 16, 2019, https://time.com/5623799/apollo-11-mission-control/.

3. "Sally Ride: Breaking Barriers, Blazing Trails," Civil Air Patrol, March 24, 2020, https://www.cap.news/sally-ride-breaking-barriers-blazing-trails/.

4. Jim Wilson, "Human Exploration in and beyond Low Earth Orbit," NASA, January 29, 2009, https://www.nasa.gov/directorates/heo/reports/iss_reports/2009/01292009.html.

5. Janet Ogle-Mater, "Sally Ride: The First American Woman in Space," ThoughtCo, last updated January 23, 2020, https://www.thoughtco.com/sally-ride-1779837.

6. Scott Kelly, *Endurance: A Year in Space, A Lifetime of Discovery* (New York: Alfred A. Knopf, 2017), 210.

7. Mary Roach, *Packing for Mars: The Curious Science of Life in the Void* (New York: WW Norton, 2011), 189–190.

8. Roach, *Packing for Mars*, 190.

9. Chris Hadfield, "Training and Learning: Simulations," Chris Hadfield Teaches Space Exploration, Masterclass, accessed on May 14, 2021, online course video, https://www.masterclass.com/classes/chris-hadfield-teaches-space-exploration/chapters/training-and-learning-simulations.

10. Hadfield, "Simulations."

11. Hadfield, "Simulations."

12. Nancy Bray, "Astronauts," NASA, April 28, 2015, https://www.nasa.gov/centers/kennedy/about/information/astronaut_faq.html#5.

13. "Astronaut Candidate," USAJOBS, accessed May 14, 2021, https://www.usajobs.gov/GetJob/ViewDetails/561186900.

14. "GS-12 Pay – 2021 Federal GS Payscale," FederalPay, accessed May 14, 2021, https://www.federalpay.org/gs/2021/GS-12.

15. Rachael Blodgett, "Frequently Asked Questions," NASA, January 16, 2018, https://www.nasa.gov/feature/frequently-asked-questions-0/.

16. Robert Frost, reply to "How much does an astronaut working on the ISS get paid?" Quora, August 21, 2013, https://qr.ae/pNgAE2.

17. "O-4 Basic Pay Rate – OFFICER Military Payscales," FederalPay, accessed September 1, 2021, https://www.federalpay.org/military/grades/o-4.

18. "O-6 Basic Pay Rate – OFFICER Military Payscales," FederalPay, accessed September 1, 2021, https://www.federalpay.org/military/grades/O-6.

19. "Frequently Asked Questions – Astronauts," Canadian Space Agency, January 9, 2020, https://www.asc-csa.gc.ca/eng/astronauts/faq.asp.

20. "Salary Table," NCIA-NATO Communications and Information Agency, January 1, 2020, https://www.ncia.nato.int/resources/site1/general/employment/what_we_offer/salary_table_20200101_website.pdf.

21. "What We Offer," European Space Agency, accessed May 14, 2021, https://www.esa.int/About_Us/Careers_at_ESA/What_we_offer.

22. Frost, "Get paid."

23. Kelly, *Endurance*, 91.

24. Economic Research Institute, "Astronaut Salary – Japan," Salary Expert, accessed May 14, 2021, https://www.salaryexpert.com/salary/job/astronaut/japan.

CIVILIAN ASTRONAUTS

By Invitation Only

U.S. SENATOR – JAKE GARN

Jake Garn was the first public official and first guest astronaut invited to go into space on the space shuttle. Garn is from Utah and chaired a NASA oversight committee. He flew as a "congressional observer" on a seven-day flight in April 1985.

Garn took part in an informal "Toys in Space" study where he and the crew demonstrated on video how a paper airplane "flies" in space, how a slinky reacts, what happens to a windup mouse and how hard it is to play jacks without gravity.[1]

SAUDI ARABIAN PRINCE – PRINCE SULTAN BIN SALMAN BIN ABDULAZIZ AL SAUD

In June of 1985—a few months after guest astronaut Jake Garn flew into space—the 28-year-old Saudi Prince Sultan bin Salman bin Abdulaziz Al Saud, a Saudi Air Force pilot, also flew into space as the second guest astronaut. He became the first Arab, Muslim and royal in space.

During his week-long mission onboard space shuttle *Discovery*, Sultan bin Salman helped to release an Arabsat communication satellite, gave a tour of the shuttle in Arabic that was broadcast on television in the Middle East and carried out experiments designed by Saudi scientists.[2]

U.S. CONGRESSMAN – BILL NELSON

Bill Nelson was the second politician and third guest astronaut invited to fly in space. He flew early in January 1986 onboard space shuttle *Columbia*. It was the last successful shuttle mission before the *Challenger* disaster a few weeks later.

NASA's Space Flight Participant Program

In the early 1980s, NASA established the Space Flight Participant Program that would send regular citizens into space. NASA wanted to not only increase public interest and support, but also send nonprofessional astronauts into space who would be able to tell the world what it was really like up there. The plan was to start with a teacher, then a journalist and then an artist.

NO SHORTAGE OF VOLUNTEERS

Once the word got out that NASA would be sending everyday citizens into space onboard the space shuttle, several people expressed interest, including singer John Denver and millionaire publisher Malcom Forbes.

BIG BIRD IN SPACE?

During the Space Flight Participant Program, NASA officials had discussions[3] about inviting the famous *Sesame Street* character Big Bird (and the puppeteer inside, Caroll Spinney) to fly onboard the space shuttle *Challenger* and teach kids about space in the process. The plan to send the famous puppet onboard was never approved, maybe because logistically it would be hard to squeeze an 8-foot-2-inch-tall puppet into the space shuttle.[4]

Teachers in Space Program

A TEACHER IN SPACE

NASA decided that a teacher would be the first civilian passenger. They sent over 40,000 applications to teachers who had expressed interest in the program. Of those, over 11,000 sent back completed applications. These were narrowed down to 114 semifinalists and included two teachers from each U.S. state plus a few more. Ten finalists were chosen which included six women and four men who trained together for a short time. Eventually, 35-year-old high school history teacher Christa McAuliffe was chosen in July of 1985.[5] Barbara Morgan was her backup.

TEACHERS TRAIN AS ASTRONAUTS

Both McAuliffe and Morgan took a year-long leave of absence so they could train for five to six months for a space shuttle mission in early 1986 and then go on tour after to share the experience. NASA paid both of their salaries while they trained. Just like other astronauts, McAuliffe and Morgan tested out the food they'd eat while in space as well as experience weightlessness onboard planes that simulate microgravity by flying roller-coaster type paths in the air. These planes and their flight path earned the nickname, "vomit comets" because they can make some passengers nauseous and lose their lunch.[6] They spent 100 hours of flight training at the NASA Johnson Space Center in Houston.[7]

LOSING MCAULIFFE & THE CREW OF THE *CHALLENGER*

Christa McAuliffe was part of the crew who died when the Space Shuttle *Challenger* broke apart 73 seconds into its flight the morning of January 28, 1986, killing everyone onboard.

LESSONS FROM LOSS

McAuliffe had planned to share six filmed lessons along with two 15-minute live lessons while in space.[8] While McAuliffe was never able to share these lessons, her lessons are still available. And several of them were brought to life by other astronauts while in space onboard the ISS.[9]

McAuliffe had named her live lessons. The first was "The Ultimate Field Trip" where she was going to give a tour of the Shuttle and introduce some of the crew. The second live lesson was called "Where We've Been and Where We're Going." These would have been broadcast on PBS during the sixth day of the mission.[10]

Videos of McAuliffe rehearsing these lessons can be found on YouTube and you can watch her in action. NASA officials described McAuliffe as having "infectious enthusiasm" and you can see that come through in these videos.

"I touch the future. I teach." **– Christa McAuliffe**

"Space is for everybody. It's not just for a few people in science or math, or for a select group of astronauts. That's our new frontier out there, and it's everybody's business to know about space."
– Christa McAuliffe, December 6, 1985

 DID YOU KNOW?
McAuliffe is a small crater on the Moon that was named in honor of Christa McAuliffe in 1988.[11]

THE END OF THE TEACHER IN SPACE PROGRAM

After the *Challenger* Disaster, NASA suspended the Teacher in Space Program indefinitely and eventually cancelled the program in 1990 saying spaceflight was too dangerous to risk the lives of civilian teachers.

BARBARA MORGAN: EXTRA SPECIAL BACKUP

Barbara Morgan was a second-grade teacher from Idaho when she was selected to be backup for Christa McAuliffe. She trained alongside McAuliffe along with the rest of the *Challenger* crew and was ready to complete the mission in case McAuliffe wouldn't be able to fly.

After McAuliffe's death, Morgan volunteered to go on the national tour that had been planned for McAuliffe after her historic mission. From March to July 1986, Morgan visited schools all over the country to talk about the space shuttle and the importance of education and space exploration. She wanted school children to hear from someone who had shared McAuliffe's dream of flying in space and to help keep their faith in the space program alive.[12] She returned to teaching in the fall of 1986.

MORGAN BECOMES AN ASTRONAUT

In 1998, twelve years after McAuliffe's death, Morgan was selected to become an astronaut candidate. She spent two years training as a full-time professional astronaut and then worked as CAPCOM (the person in Mission Control responsible for speaking to the astronauts in space) in mission control for a time. She flew into space onboard *Endeavour* in 2007 on STS 118 as a Mission Specialist (not an Educator Astronaut or Mission Specialist Educator as she is often wrongly labelled). She spent 12 days in space and was the space shuttle robotic arm operator that transferred over 5,000 lbs (2,300 kg) of cargo to the International Space Station.[13]

MORGAN TALKS TO STUDENTS FROM THE ISS

Morgan is a licensed amateur radio operator and participated in a 20-minute question and answer session with youth at Discovery Centers in Idaho while onboard the International Space Station.[14] Read and listen to that conversation in the NASA archives.

"Reach for your dreams ... the sky is no limit."
– Barbara Morgan

 DID YOU KNOW?

YOU CAN TALK TO ASTRONAUTS ON THE ISS VIA HAM RADIO

Thanks to the Amateur Radio on the International Space Station (ARISS) program, there's a ham radio onboard the space station and crew members chat with groups of people (usually students) from around the world about 45 times a year.[15]

Astronauts have also made random, unscheduled, amateur radio voice contacts with amateur radio operators, often called "hams," during their free time in space.[16]

To learn which astronauts are licensed hams, what ISS call signs are in use and how you can chat with astronauts in space, please refer to the resources in the back of this book.

Educator Astronaut Project

Following the *Challenger* Disaster, the Teacher in Space Program was cancelled. In 1998, a new program was created called the Educator Astronaut Project. It was designed to inspire students and foster excitement in learning math, science and space exploration. Rather than send civilians who were teachers into space, NASA selected applicants who were former teachers to become astronauts. They would be fully trained astronauts with the role of Mission Specialists-Educators.

The first Educator Astronauts were part of the Astronaut Candidate Class of 2004 and included these three classroom teachers:

JOE ACABA

Acaba was 36 years old when he was selected as an Astronaut Candidate in 2004. He became the first person of Puerto Rican heritage to be named a NASA astronaut. Before becoming an astronaut, he was a teacher in Florida and a Peace Corps volunteer in the Dominican Republic. Since then, he's been to space three times and has spent a total of 306 days in space, mostly onboard the International Space Station.

RICKY ARNOLD

When he was selected, Arnold was a 40-year-old math and science teacher who had taught science and math in Maryland, Saudi Arabia, Romania, Morocco and Indonesia. He went to space twice, once on the space shuttle and once on a Soyuz spacecraft, spending time onboard the ISS. He performed five EVAs and spent a total of 209 days in space.

DOTTIE METCALF-LINDENBURGER

Metcalf-Lindenburger was 29 years old and teaching high school science in Vancouver, Washington, when she was selected as an Astronaut Candidate in 2004. She was also the first Space Camp alumni to become an astronaut. She's been to space once onboard the space shuttle *Discovery*. She spent 15 days in space.

Journalist in Space Program

To help communicate what space was like in an understandable way, NASA had plans to send a journalist into space in September of 1986 after they sent a teacher.

OVER 1,700 APPLICANTS

Walter Cronkite from CBS, Tom Brokaw with NBC and Sam Donaldson from ABC were among the 1,703 applicants who wanted to be the first journalist in space.

AFTER *CHALLENGER* DISASTER

The Journalist in Space program was postponed after the *Challenger* disaster, but none of the 1,703 applicants withdrew their applications. Some even called NASA to make sure they knew they still wanted to go to space.[17]

40 SEMI-FINALISTS

The selection process still continued even though the program was put on hold. From the original 1,703 applicants, 100 region finalists were chosen. Out of the 100, 40 semi-finalists were chosen. The list of 40 included a variety of journalists from different backgrounds. There were 15 newspaper writers, 12 people from television, two from radio, three from magazines, three from wire services and five freelance journalists. Walter Cronkite was among those finalists.[18]

POSTPONED INDEFINITELY

NASA's Journalist in Space program was eventually postponed indefinitely in July of 1986.[19]

Business Class to Space

The first non-professional astronauts weren't space tourists, they were people who were on a business trip of sorts. The Russian Space Program has sometimes struggled with funding and they learned that there are companies and people who will pay to get a ride to space on their Soyuz rocket. It's a win/win for both parties. Russia gets the cash they need and the lucky participant gets to fly with professional astronauts into space.

FIRST NONPROFESSIONAL ASTRONAUT IN SPACE
The first journalist and first nonprofessional astronaut sent to space was from Japan. Toyohiro Akiyama launched onboard a Soyuz rocket as part of a $12 million deal with the Soviet Union and the Tokyo Broadcasting System in 1990. Akiyama was a Japanese TV journalist and was also the first Japanese citizen sent into space. The mission was called Mir Kosmoreporter, and Akiyama spent 8 days onboard the space station Mir, where he made daily television broadcasts.[20]

FIRST BRITISH ASTRONAUT
Helen Sharman worked as a chemist at Mars, the chocolate bar company, when she heard a radio ad calling for applicants to be the first British Space Explorer. Called Project Juno, this program was a joint project between private British companies and the Soviet Union. Roughly 13,000 people applied and 26-year-old Sharman was chosen to be the first British astronaut.

Sharman was required to learn Russian before she started her intensive space training. She spent 18 months training for her mission in Star City, Russia, where she became part of the cosmonaut community and learned everything she would need to know for her mission to space and onboard the space station Mir.[21]

On May 18, 1991, Sharman and two Russian cosmonauts launched on the Soyuz TM-12. She was the first woman to visit the space station Mir and while she was there, Sharman conducted medical, chemical and agricultural experiments, spoke to British schoolchildren over amateur radio and observed and photographed parts of the UK from space.[22]

Space Tourists
They prefer to call themselves spaceflight participants, but to the

world they are known as space tourists. These seven men and women were the first to personally pay their way to space onboard a Russian Soyuz and then stay on the International Space Station. They trained alongside cosmonauts for 8 months or more and learned Russian to prepare for their out-of-this-world experience.

#1 AMERICAN BUSINESSMAN — DENNIS TITO

On April 28, 2001 Dennis Tito became the world's first space tourist. He paid about $20 million for a seat on a Russian Soyuz bound for the International Space Station. After 8 months of training in Star City, Russia, he spent 6 days onboard the ISS. While he mostly stayed on the Russian side, he did get a special tour of the U.S. side. With launch and re-entry, his entire trip took just under 8 days—so he paid over $2 million per day in space![23]

#2 SOUTH AFRICAN INTERNET MILLIONAIRE – MARK SHUTTLEWORTH

In April 2002, Mark Shuttleworth became the first African in space and the second space tourist at the age of 28. He reportedly also paid about $20 million and spent a year training—with 8 months of it in Star City, Russia. He spent a total of 8 days on the ISS. During his time there, Shuttleworth conducted experiments related to AIDS and genome research. While in space, he was able to call and speak to Nelson Mandela and Thabo Mbeki, the president of South Africa at the time.[24]

#3 AMERICAN ENTREPRENEUR & SCIENTIST – GREG OLSEN

The third person to fly into space as a "spaceflight participant" was Greg Olsen. He almost missed out on the trip. While training for his flight in Russia, a black spot was discovered on his lung during a routine medical exam. They sent him home and said he was unable to fly. But he didn't give up!

The black spot eventually went away, and Olsen applied 8 more times—and was rejected all 8 times. It wasn't until 9 months later with a note from his doctor and after signing a thick stack of release forms that he was allowed to fly.[25] Olsen finished his training, was healthy, and launched into space October 1, 2005. While onboard the ISS, he conducted several experiments related to astronomy and remote sensing. Olsen is a licensed Amateur Radio operator and was able to speak to students back on Earth via ham radio. He returned home 10 days later. His trip reportedly also cost around $20 million.

#4 TECH BUSINESSWOMAN – ANOUSHEH ANSARI

Not only did Ansari become the first female space tourist, but she was also the first Muslim woman and first Iranian in space. Ansari reportedly paid around $20 million for her ticket to space. She originally was the backup for space tourist, Daisuke Enomoto. They were training together in Star City, Russia. Due to medical reasons, Enomoto was unable to fly and Ansari replaced him. She flew onboard Soyuz TMA-9 on September 18, 2006.[26]

Ansari spent 8 days onboard the ISS participating in psychology experiments for the European Space Agency. She was also interviewed by an astronomy show that aired on Iranian National TV and shared her experience on her blog.[27] Ansari was actually the first person to blog while in space! She returned home on September 29, 2006. She has written a book about her experience, *My Dream of Stars: From Daughter of Iran to Space Pioneer*.

Ansari and her brother-in-law, Amir Ansari, sponsored the space competition called the Ansari X Prize. The Ansari X Prize offered 10 million US dollars, the largest prize in history, to the first privately funded organization to launch a reliable, reusable and crewed spacecraft into space twice within two weeks. The first Ansari X Prize was awarded in October 2004 to the Mojave Aerospace Ventures for spaceplane, SpaceShipOne. Sir Richard Branson licensed this award-winning technology to create Virgin Galactic. The Ansari X Prize competition led to the launch of a new, $2 billion private space industry.[28]

#5 MICROSOFT OFFICE DEVELOPER — CHARLES SIMONYI

Few people on Earth get the chance to pay their way to space onboard a Russian Soyuz. Only one person in history has paid their way to space twice. Hungarian-born Charles Simonyi paid close to $60 million for two roundtrip tickets to the International Space Station, first in 2005 and the second time in 2007. He is the second Hungarian to have been to space. Due to delays coming home due to weather on Earth, he spent a space tourist record of 14 days in space during his first trip.

#6 GAMING TITAN & COFOUNDER OF SPACE ADVENTURES — RICHARD GARRIOTT

It's not every day that a former astronaut has a son that gets to follow in his footsteps. Astronaut Owen Garriott was one of NASA's first scientist astronauts. He spent 60 days onboard the Skylab space station and 10 days on Space Shuttle *Columbia*. His son, Richard, wanted to become an astronaut, but issues with his eyesight that began at the age of 13 disqualified him. So, Richard decided to figure out his own way to space—and he did!

Richard cofounded Space Adventures, the only company which has made it possible to send paying private civilians into space in partnership with the Russian space program. He was meant to be the first space tourist, but after the dot-com bubble burst, he faced financial difficulties and sold his seat to Dennis Tito. He was able to regain his fortune and bought a second ticket to space, reportedly around $30 million. He launched into space on October 12, 2008 and returned home 12 days later.[29]

During his 12-day trip to the International Space Station, Richard secretly stashed the ashes of *Star Trek* actor James Doohan (who played Scotty) under the floor in the ISS's Columbus module. The ashes were contained in a laminated card. This wasn't discovered until December 2020, when Doohan's son made an announcement on Twitter. More than a decade later, Doohan's ashes have travelled more than 1.7 billion miles and orbited Earth more than 70,000 times.[30]

DID YOU KNOW?

THERE'S A GEOCACHE ON THE ISS

A geocache is a hidden "treasure" that can be found using a GPS device or mobile app.

During his visit to the ISS, Richard Garriott left what is most likely the only out-of-this-world geocache onboard the International Space Station.

As of September 1, 2021, only two people have found it.[31]

#7 CIRQUE DU SOLEIL FOUNDER — GUY LALIBERTÉ

Guy Laliberté is the most recent space tourist, and also the first Canadian private citizen to pay his way to space. He paid between $35 and $41 million for his trip to the ISS, which launched on September 30, 2009 and returned home 11 days later. During his "poetic social mission," he hosted a 2-hour-long live event from space with other stars in 14 different cities. It was all about raising awareness for the One Drop Foundation, which gives access to safe water, sanitation and hygiene to those who need it worldwide.[32] Laliberté took hundreds of photos while in space.[33] He later published the book *Gaia* with many of his photos of Earth. All proceeds go to the One Drop Foundation.

Inspiration4 — The First SpaceX Space Tourists

SpaceX flew four space tourists on the SpaceX Crew Dragon *Resilience* capsule in September 2021. The entire flight was bought by billionaire Jared Isaacman for an undisclosed amount. For a ballpark figure, future SpaceX space tourist flights currently list tickets at roughly $55 million each for a three-to-five-day trip to space.

THE PRIVATE CITIZEN CREW

Isaacman is a highly experienced pilot and acted as the spaceflight commander. This historic flight was dedicated to fundraising money for

St. Jude's Children Research Hospital to help tackle childhood cancer. Isaacman chose three other crew members to help represent and symbolize specific attributes related to the mission: hope, generosity and prosperity. Isaacman as the commander, symbolized leadership.

JARED ISAACMAN

Billionaire Jared Isaacman is a tech entrepreneur and record-setting pilot. He founded a retail payment processing company similar to PayPal, called Shift4 Payments and processes more than $200 billion in payments for restaurants and hotels like Hilton and KFC. He co-founded Draken International, which provides tactical aircraft, like fighter jets and trains pilots for the U.S. military. Isaacman is a pilot and set a world speed record flying around the world in 2009 to raise money for the Make-A-Wish Foundation. He's wanted to go to space since he was a young boy and jumped at the chance when SpaceX made it a possibility.[34]

SIAN PROCTOR

Dr. Sian Proctor is a geoscientist, explorer, entrepreneur and represented the pillar of prosperity. Before she made it to space, she was an analog astronaut and adventure scientist. She completed four analog missions, including the NASA funded four-month Mars mission at Hi-SEAS (Hawai'i Space Exploration Analog and Simulation) and a two-week mission at the Mars Desert Research Station (MDRS). Proctor has applied to become a NASA astronaut three separate times. In 2009, she was among 47 semifinalists, but was not part of the final nine chosen.[35] She finally got her chance to fly into space, when Proctor was selected as the pilot for the Inspiration4 mission.[36]

HAYLEY ARCENEAUS

Hayley Arceneaus is a bone cancer survivor and front-line worker from St. Jude Children's Research Hospital and represented the pillar of hope on the Inspiration4 flight. She was a St. Jude Children's Research Hospital patient at the age of 10 and is now a physician assistant at St. Jude. Arceneaus became the youngest American in space at 29 when she flew into space in September 2021.

CHRISTOPHER SEMBROSKI

Christopher Sembroski is a data engineer and Air Force veteran who has a deep interest in space and astronomy. While Sembroski donated and participated in a raffle to help raise $200 million for St. Jude's and win a seat on Inspiration4, his raffle ticket was not selected. A friend of his who has chosen to remain unnamed had their ticket chosen to fly with Isaacman, but he declined for personal reasons and transferred his seat to Sembroski. Sembroski represented the pillar of generosity.

COMMERCIAL ASTRONAUT TRAINING

All four crew members spent several months undergoing commercial astronaut training by SpaceX to prepare for their flight. Training included emergency preparedness, spacesuit exercises and partial and full mission simulations. They also focused on orbital mechanics and how to operate in microgravity. SpaceX based most of its training curriculum on the program NASA uses for its astronauts.[37]

ORBIT-ONLY

This flight did not include a visit to the International Space Station. The Inspiration4 flight was an orbit-only trip, but a very high one. Inspiration4 flew to the planned altitude of 364 miles (585 km)— higher than any crewed orbital spacecraft before! For reference, the ISS flies at an average altitude of 248 miles (400 km) above Earth.

MAKING HISTORY

Inspiration4 was the first human spaceflight to orbit Earth with only private citizens onboard. They circled the Earth for three days and enjoyed an epic view of the Earth. SpaceX installed a large, bubble-like dome window that allowed a 360-degree view and gave a better view than most astronauts have had in the 60-year history of spaceflight. The trip started at the Kennedy Space Center and ended with a splashdown in the Atlantic Ocean.

Future Space Tourists

POSTPONED: *THE PHANTOM OF THE OPERA STAR* — SARAH BRIGHTMAN

The best-selling British singer, Sarah Brightman, was set to be the 8th space tourist in September 2015, but she postponed her trip suddenly in the middle of training in Star City, Russia. She announced on her blog that it was due to "personal family reasons." She reportedly paid around $50 million for the experience. As of April 2021, she has not rescheduled her space flight.[38]

ON DECK: ADVERTISING ENTREPRENUER — SATOSHI TAKAMATSU

Takamatsu began training in 2015 in Star City Russia as Sarah Brightman's backup. When Brightman canceled her trip, Satoshi declined to take her spot. His plan in space involves several art projects meant to be carried out in space. These projects weren't ready to go into space by the September 15th date so he asked to schedule his flight for a later date. As of September 2021, he is still waiting for a seat on the Soyuz to open up to take him to the ISS.

MISSED OPPORTUNITY: BUSINESSMAN — DAISUKE ENOMOTO

Enomoto was meant to be the 4th space tourist, but was medically disqualified to fly due to kidney stones shortly before launch. His backup, Anousheh Ansari, flew in his place. He has sued Space Adventures in hopes of getting his $21 million back.[39]

2021 PLANNED TRIPS

Roscosmos has partnered with Channel One, a Russian television network, to send two people to the ISS for 12 days in October 2021. Actress Yulia Peresild and director Klim Shipenko will join Russian cosmonaut Anton Shkaplerov on the Soyuz -MS-19 mission. During the 12 days, Peresild and Shipenko will shoot scenes for a Russian movie called "Vyzov" (The Challenge).

Space Adventures and the Russian Space Agency are working together again and will fly two space tourists to the International

Space Station in December 2021. The Japanese billionaire Yusaku Maezawa, the man who bought an entire SpaceX Starship flight around the Moon in 2023 and is inviting eight artists to join him in a project known as dearMoon, purchased two seats. Maezawa and his production assistant, Yozo Hirano, will fly to the ISS on the Soyuz MS-20 mission with Russian cosmonaut Alexander Misurkin on a 12-day trip. Maezawa and Hirano will become the first space tourists from Japan.[40]

Space Tourism Companies

SPACE ADVENTURES

As of April 2021, Space Adventures is the only space tourism company to have sent private citizens that pay their own way into space. They have coordinated eight trips for seven space tourists.

Until recently, Space Adventures has worked exclusively with the Russian space program to send all seven space tourists on a Soyuz spacecraft headed to the International Space Station. After the Space Shuttle program was retired, Russia was the only country sending people into space. During that time, Space Adventures had to compete with NASA for seats on the Soyuz. But now that NASA is using SpaceX's Crew Dragon to take its astronauts into space, seats should become more available.

Now that private companies outside of government agencies are working to send humans into space, Space Adventures has begun working with them. Space Adventures has added additional space adventure opportunities to their offerings, including five-day suborbital flights, flights around the Moon, spacewalks and more flights to the International Space Station.[41]

ELON MUSK'S SPACEX

The same Falcon rocket and Crew Dragon capsule that has been shipping cargo—and most recently crew—to the International Space Station, will also be used to send four paying passengers into low earth orbit. These trips will last up to five days, and tourists will fly several times higher than the ISS. Each ticket will include a few weeks of pre-flight training in the U.S. Passengers will fly on the first Crew Dragon free-flyer mission. This spacecraft will fly on autopilot with no trained astronauts onboard, but participants will be trained to handle a variety of emergencies. SpaceX and Space Adventures are currently taking bookings and have a flight scheduled for early 2022 from Cape Canaveral in Florida.[42] Prices are roughly $55 million per person.[43]

SpaceX and Space Adventures are working together to send paying customers around the Moon in a Circumlunar mission.[44] Japanese billionaire and fashion designer Yusaku Maexawa is already booked to take SpaceX's first private space flight around the Moon onboard SpaceX's Starship. He's expected to launch in 2023. In March 2021, Maexawa announced a new project called dearMoon. He has purchased all of the seats on this first private flight around the Moon and wants to take 8 artists with him. People from around the world have applied to the dearMoon project and selection is underway.[45]

AXIOM AND SPACEX

Axiom has partnered with SpaceX to send paying passengers to the International Space Station. The first trip will send three paying guests with one company-trained astronaut who will serve as the flight commander. These 10-day trips are expected to cost $55 million per seat.[46] The first flight is scheduled for early 2022 with Eytan Stibbe from Israel, Larry Connor from the U.S. and Mark Pathy from Canada with retired NASA astronaut Michael Lopez-Alegria as the commander.

Actor Tom Cruise and film director, Doug Liman, are scheduled to fly on a future Axiom flight for a movie project.[47]

JEFF BEZOS'S BLUE ORIGIN

Amazon founder Jeff Bezos is selling tickets for suborbital space flights onboard Blue Origin's New Shepard spacecraft. Tickets for an 11-minute suborbital space flight cost between $200,000 and $300,000 per ticket.[48]

The New Shepard has six seats and six observation windows. The experience includes flying up to six passengers more than 62 miles (100 km) above Earth. This is high enough to experience a few minutes of weightlessness and to see the curvature of the Earth. The capsule will then safely return to Earth using parachutes.

In June 2021, Jeff Bezos announced that he and his brother, Mark, would be on the first New Shepard flight in July 2021. Bezos auctioned off the third spot for $28 million! [49]

Blue Origin's New Shepard sent its first crewed mission into space on July 20, 2021. The first four Blue Origin passengers were Jeff Bezos, his brother Mark, Wally Funk and Oliver Daemen. Wally Funk is one of the last two surviving members of the Mercury 13 group and the only one to have traveled to space. She set the record to become the oldest woman to travel to space. At the age of 18, Oliver Daemen set the record to become the only teenager and the youngest person to travel to space.

On October 13, 2021, New Shepard sent its second crewed mission into space and this time, William Shatner, the actor that played Captain Kirk in *Star Trek* TV series, was onboard. At the age of 90, Shatner set a new record as the oldest person to travel to space.

SIR RICHARD BRANSON'S VIRGIN GALACTIC

Before its first crewed flight in July 2021, Virgin Galactic SpaceShipTwo tickets cost about $250,000 per ticket and 600 tickets had already been sold.[50] Among the 600 ticket holders are Justin Bieber and Leonardo DiCaprio.[51] As of October 2021, tickets now cost $450,000 per seat.

Before Virgin Galactic certified their flights as public ready, Sir Richard Branson himself flew onboard to test it out first. He successfully flew along with three other Virgin Galactic employees. Ticketed passengers are expected to start flying at the beginning of 2022 after a few more test flights.

The Virgin Galactic experience will take eight people (two pilots and six paying customers) up in the SpaceShipTwo VSS Unity spacecraft. The flight duration is expected to be two to three hours with time in weightlessness around four minutes. Three days of pre-flight training and medical checkups are also included to ensure the participants are mentally and physically prepared for this life-changing experience.[52]

SPACE PERSPECTIVE HIGH ALTITUDE BALLOON FLIGHTS

For those who want a more relaxed experience and prefer not to be strapped to some sort of rocket, Space Perspective will be taking up to 12 people to the edge of space at 100,000 feet (30.4 km) for a 6-hour suborbital journey.

The roomy Spaceship Neptune space balloon includes a bar, Wi-Fi and bathroom. The space balloon is shaped like a top, which gives a 360-degree panoramic view with gold-tinted windows to give shade and protect from the sun's radiation.[53] The company plans to fly up to 100 flights per year and will offer tickets about half the price of Virgin Galactic's $250,000 per seat. Final pricing and the ability to start booking should be available later in 2021.[54]

BOEING'S STARLINER

NASA chose SpaceX and Boeing to be their official partners in shuttling astronauts to and from the ISS back in 2014. While SpaceX has been successful with their Falcon 9 rockets and Crew Dragon capsules, Boeing is still working out a few bugs with their Starliner spacecraft. Once it is fixed, they hope to also join the space tourism game. They've even hired test-pilot astronaut Christopher Ferguson, who is currently training alongside NASA's astronauts. He will be among the first to fly the Starliner spacecraft. This could give new meaning to the title "professional astronaut" and could allow them to fly for private companies.[55]

NOTES

1. "Toys in Space," UPI, accessed May 14, 2021, https://www.upi.com/Archives/1985/04/15/Toys-in-space/6791482389200/.

2. "Sultan ibn Salman Al Saud," Encyclopedia Britannica, June 23, 2020, https://www.britannica.com/biography/Sultan-ibn-Salman-Al-Saud.

3. Alan Boyle, "NASA Confirms Talks to Fly Big Bird on Doomed Shuttle Challenger," NBCNews.com, May 4, 2015, https://www.nbcnews.com/science/weird-science/nasa-confirms-talks-fly-big-bird-doomed-shuttle-challenger-n353521.

4. Erin Blakemore, "Big Bird Nearly Rode on the Disastrous Challenger Mission," History.com, January 26, 2018, https://www.history.com/news/big-bird-challenger-disaster-nasa-sesame-street.

5. Greg Daugherty, "The Challenger Disaster Put an End to NASA's Plan to Send Civilians into Space," Smithsonian Institution, January 27, 2016, https://www.smithsonianmag.com/history/challenger-disaster-put-end-nasas-plans-send-civilians-space-180957922/.

6. Nola Taylor Redd, "Vomit Comet: Training Flights for Astronauts," Space.com, August 25, 2017, https://www.space.com/37942-vomit-comet.html.

7. "Teachers in Space: A Chronology," Education Week, February 24, 2019, https://www.edweek.org/education/teachers-in-space-a-chronology/1998/01.

8. "Christa's Lost Lessons," Challenger Center, April 29, 2021, https://www.challenger.org/challenger_lessons/christas-lost-lessons/.

9. "Christa's Lost Lessons."

10. "Christa's Lost Lessons."

11. "Planetary Names: Crater, Craters: McAuliffe on Moon," USGS Astrogeology Science Center, accessed May 14, 2021, https://planetarynames.wr.usgs.gov/Feature/3777.

12. Scott Kelly, *Endurance: A Year in Space, a Lifetime of Discovery* (New York: Alfred A. Knopf, 2017), 257.

13. Brian Dunbar, "STS-118 MCC Status Report #01," NASA, August 8, 2007, https://www.nasa.gov/mission_pages/shuttle/shuttlemissions/sts118/news/STS-118-01.html.

14. Brian Dunbar, "Barbara Morgan Talks with Students on Ham Radio," NASA, August 16, 2007, https://www.nasa.gov/multimedia/podcasting/STS118_morgan_radio.html.

15. Natalie Wolchover, "How to Call Space Station Astronauts on the Radio," Livescience.com, August 18, 2011, https://www.livescience.com/33453-iss-astronauts-ham-radio.html.

16. Samantha Masunaga, "Earthlings and astronauts chat away, via ham radio," Phys.org, December 23, 2020, https://phys.org/news/2020-12-earthlings-astronauts-chat-ham-radio.html.

17. Thomas Rosenstiel, "Journalist-in-Space Plan Postponed Indefinitely," *Los Angeles Times*, January 31, 1986, https://www.latimes.com/archives/la-xpm-1986-01-31-mn-2634-story.html.

18. Joachim Becker, "Candidates for the 'Journalist in Space Program,'" Spacefacts.de, last updated March 1, 2021, http://www.spacefacts.de/english/e_journalist.htm.

19. "Journalist-in-Space Program on Hold," UPI, July 15, 1986, https://www.upi.com/Archives/1986/07/15/Journalist-in-space-program-on-hold/7053521784000/.

20. "Akiyama Toyohiro," Encyclopedia Britannica, July 18, 2020, https://www.britannica.com/biography/Akiyama-Toyohiro.

21. "About Helen," Helen Sharman CMG OBE: The First British Astronaut, accessed August 11, 2021, https://www.helensharman.uk/about-helen/.

22. Emma Doughty, "Project Juno: Pansy Seeds in Space," *The Unconventional Gardener* (blog), January 26, 2020, https://theunconventionalgardener.com/blog/project-juno-pansy-seeds-in-space/.

23. "Dennis Tito," Encyclopedia Britannica, August 4, 2020, https://www.britannica.com/biography/Dennis-Tito.

24. "Mark Shuttleworth," Space Adventures, January 12, 2018, https://spaceadventures.com/mark-shuttleworth/.

25. Jim Clash, "How Greg Olsen Got a Bargain when He Spent $20 Million of His Own Money to Fly in Space," Forbes, February 23, 2017, https://www.forbes.com/sites/

jimclash/2017/02/23/how-greg-olsen-got-a-bargain-when-he-spent-20-million-of-his-own-money-to-fly-in-space/?sh=2408ad815d12.

26. "Anousheh Ansari," Encyclopedia Britannica, September 8, 2020, https://www.britannica.com/biography/Anousheh-Ansari.

27. Anousheh Ansari, "Anousheh's Space Blog," accessed May 14, 2021, http://www.anoushehansari.com/blog/.

28. "Launching a New Space Industry," XPRIZE, accessed September 2, 2021, https://www.xprize.org/prizes/ansari.

29. Catherine Clifford, "What It's like to Travel to Space, from a Tourist Who Spent $30 Million to Live There for 12 Days," CNBC, October 19, 2018, https://www.cnbc.com/2018/10/19/what-its-like-in-space-from-a-tourist-who-spent-30-million-to-go.html.

30. Jacqui Goddard, "Ashes of Star Trek's Scotty Smuggled on to International Space Station," The Times, December 25, 2020), https://www.thetimes.co.uk/article/ashes-of-star-treks-scotty-smuggled-on-to-international-space-station-6lpgs05n6.

31. Richard Garriott, "GC1BE91 International Space Station," Geocaching.com, October 14, 2008, https://www.geocaching.com/geocache/GC1BE91_international-space-station.

32. "Guy Laliberte," Space Adventures, January 12, 2018, https://spaceadventures.com/guy-laliberte/.

33. Steven Bertoni, "Why Cirque Du Soleil Billionaire Guy Laliberte Traveled to Space," Forbes, August 11, 2011, https://www.forbes.com/sites/stevenbertoni/2011/06/09/why-cirque-du-soleil-billionaire-guy-laliberte-traveled-to-space/?sh=7fa5156d7cb2.

34. Passant Rabie, "Who Is Jared Isaacman? 29 Facts about the Billionaire Going to Orbit with SpaceX," Inverse, February 5, 2021, https://www.inverse.com/science/jared-isaacman-spacex-codex.

35. "One of NASA's Solar System Ambassadors Will Soon Be an Astronaut," NASA, June 21, 2021, https://solarsystem.nasa.gov/news/1898/one-of-nasas-solar-system-ambassadors-will-soon-be-an-astronaut/.

36. "Dr. Sian 'Leo' Proctor," accessed September 6, 2021, http://www.drsianproctor.com/.

37. Thomas Burghardt, "SpaceX Announces Inspiration4, ALL-CIVILIAN Space Mission in Support of St Jude's Hospital," NASASpaceFlight.com, February 1, 2021, https://www.nasaspaceflight.com/2021/02/spacex-announces-inspiration4/.

38. Jonathan Amos, "Sarah Brightman Calls off Space Trip," BBC News, May 14, 2015, https://www.bbc.com/news/science-environment-32733134.

39. Becky Iannotta, "Grounded Space Tourist Sues for $21 Million Refund," Space.com, October 1, 2008, https://www.space.com/5845-grounded-space-tourist-sues-21-million-refund.html.

40. Jeff Foust, "Japanese Billionaire, Russian Actress to Fly to ISS," SpaceNews, May 14, 2021, https://spacenews.com/japanese-billionaire-russian-actress-to-fly-to-iss/.

41. Space Adventures, May 13, 2021, https://spaceadventures.com/.

42. Kenneth Chang, "There Are 2 Seats Left for this Trip to the International Space Station," New York Times, March 5, 2020, https://www.nytimes.com/2020/03/05/science/axiom-space-station.html.

43. "Space Adventures, SpaceX to Launch Private Crew to Gemini Heights," collectSPACE.com, February 18, 2020, http://www.collectspace.com/news/news-021820a-space-adventures-spacex-crew-dragon.html.

44. Tom Shelley, "Space Adventures Announces Agreement with SpaceX to Launch Private Citizens on the Crew Dragon Spacecraft," Space Adventures, February 18, 2020, https://spaceadventures.com/experiences/low_earth_orbit/.

45. "8 Crew Members Wanted!" dearMoon, accessed September 2, 2021, https://dearmoon.earth/.

46. Chang, "Two Seats Left."

47. Robert Z. Pearlman, "Axiom Space Names First Private Crew to Visit Space Station," *Scientific American*, January 26, 2021, https://www.scientificamerican.com/article/axiom-space-names-first-private-crew-to-visit-space-station/.

48. Eric M. Johnson, "Exclusive: Jeff Bezos Plans to Charge at Least $200,000 for Space Rides – Sources," Reuters, July 12, 2018, https://www.reuters.com/article/us-space-blueorigin-exclusive/exclusive-jeff-bezos-plans-to-charge-at-least-200000-for-space-rides-sources-idUSKBN1K301R.

49. Michelle Acevedo and Dennis Romero, "Seat on Bezos' Blue Origin Flight to Space Sells for $28 Million at Auction," NBC News, June 12, 2021, https://www.nbcnews.com/news/us-news/seat-bezos-blue-origin-flight-space-sells-28-billion-auction-n1270595.

50. Sam Meredith, "Richard Branson Says Virgin Galactic Will Be in Space for Test Flights 'in Weeks Not Months,'" CNBC, https://www.cnbc.com/2018/10/09/richard-branson-space-virgin-galactic-weeks-not-months.html.

51. Jonathan Amos and Victoria Gill, "Space Tourism: Virgin Space Plane to Fly above New Base," BBC News, https://www.bbc.com/news/science-environment-55279067.

52. "Learn," Virgin Galactic, accessed May 14, 2021, https://www.virgingalactic.com/learn/.

53. "FLY," Space Perspective, April 28, 2021, https://thespaceperspective.com/fly/.

54. Loren Grush, "New Company Space Perspective Wants to Take You to the Stratosphere via High-Altitude Balloon," The Verge, June 18, 2020, https://www.theverge.com/21294039/space-perspective-stratosphere-balloon-travel-tourism-world-view.

55. Eric Betz, "Six Ways to Buy a Ticket to Space in 2021," Astronomy.com, August 28, 2020, https://astronomy.com/news/2020/08/six-ways-to-buy-a-ticket-to-space-in-2021.

CREWED MISSIONS

There have been dozens of different space programs, even more missions and hundreds of spacecrafts (crewed and uncrewed). Because this book is all about astronauts, the focus will be on the crewed space missions and the spaceships they flew on.

NASA's Project Mercury (1959-1963)

The Mercury program was named after the Roman god of speed. Astronaut Scott Carpenter said, "In those days, speed was magic... and nobody had gone that fast."[1] The seven astronauts who flew on the Mercury space missions were called the Mercury 7. Each astronaut named their spacecraft and all of them ended the name with the number "7."

MERCURY WAS NOT THE ORIGINAL NAME

The first NASA program to send humans into space was originally named "Project Astronaut." The U.S. President at the time, Dwight D. Eisenhower, felt that put too much of the spotlight on the astronauts. So the name was changed to Project Mercury instead.[2]

MERCURY 7 ASTRONAUTS

The seven original American astronauts were required to have a bachelor's degree (or equivalent) in engineering, be a qualified jet pilot, have graduated from test pilot school with at least 1,500 hours of flying time. They were also required to be in excellent physical condition, no more than 5 feet 11 inches (180 cm) tall and weigh no more than 180 lbs (82 kg), because of the very tight conditions inside the Mercury spacecrafts.[3] While there were seven Mercury astronauts, there were only six Mercury missions since Deke Slayton was grounded for medical reasons.

Freedom 7: Alan Shepard – First American In Space

Shepard made history as the first American in space when he flew onboard the Freedom 7, also known as the Mercury-Redstone 3 flight. He reached a maximum of 6.3 g's as he rode into space.[4] G in this book refers to a unit of measurement for the perceived weight felt in a spacecraft. For reference, 1 g is equal to a person's or object's normal weight on the Earth's surface. Astronauts on the Space Shuttle normally reached 3 g's, while the Russian Soyuz and SpaceX flights normally reach 4 g's during launch.[5]

Shepard also walked on the Moon in 1971, and hit two golf balls while he was there. He started the tradition of naming the Mercury capsules when he named his Freedom 7.

Liberty Bell 7: Gus Grissom – Second American In Space

Grissom was the second American in space. Grissom named his capsule Liberty Bell because it had a white, irregular paint stripe that started 1/3 of the way from the bottom of the capsule and continued to the bottom, which looked like the crack in the famous Liberty Bell. After splashdown, Liberty Bell 7's hatch was blown off and the capsule sank. Gus was able to exit safely and was rescued soon after.

Friendship 7: John Glenn – First American To Orbit Earth

John Glenn was the first American to orbit earth, onboard Friendship 7. He was also the first to fly both on a Mercury mission and a Space Shuttle mission and he set the record for oldest person in space when he flew onboard the space shuttle at the age of 77. Glenn's children helped him come up with the name of "Friendship" for his capsule.[6]

Aurora 7: Scott Carpenter – Second American To Orbit Earth

Scott Carpenter flew in Deke Slayton's place after Slayton was grounded for having abnormal heart rhythms. Slayton would have named the capsule Delta 7, but instead Scott named it Aurora 7 for the open sky and the dawn to symbolize the dawn of a new day.

Carpenter was the second American to orbit Earth, the first to eat solid food and was able to identify the mysterious "fireflies" that John Glenn had seen on the flight before him. He renamed them "frostflies" since they were the result of frozen condensation that would break off the capsule and become illuminated by the sun.

Sigma 7: Wally Schirra – Fifth American In Space

Schirra was the first astronaut to fly into space three times and is the only American to have flown on Mercury, Gemini and Apollo missions. Sigma 7's mission was more about the operation of the spacecraft than it was about scientific experimentation. The name Sigma was meant to symbolize this focus.

Faith 7: Gordon Cooper – Last Mercury & Solo American Orbital Mission

Cooper was the first American to spend an entire day in space, first to sleep in space and spent more time in space than all of the other American astronauts before him with 34 hours, 19 minutes and 49 seconds. He named his spacecraft Faith 7 because of his faith in the Atlas booster and Mercury spacecraft.

Delta 7: Deke Slayton – Grounded Astronaut

Slayton[7] was supposed to be the fourth American to fly into space onboard the capsule he would have named Delta 7. Delta is the fourth letter in the Greek alphabet and this would have been the fourth crewed American space flight. Slayton was grounded due to abnormal heart rhythms discovered during training and did not fly and there was no Delta 7. Scott Carpenter flew in his place in the capsule he named Aurora 7.

While Slayton was one of the original Mercury 7 astronauts chosen in 1959, he was the only one who didn't fly a Mercury mission. It took him a long 16 years before he was medically qualified to fly to space in 1975 during the Apollo-Soyuz test mission. He was 51 years old at the time of his flight and set the record for being the oldest astronaut to fly to space at that time.

NASA's Gemini Program (1961-1966)

The word gemini is Latin for twins. The Gemini capsules could hold two people instead of the single-person capsules of the Mercury mission.[8] Gemini was the program NASA used to learn everything they needed to know about spaceflight before sending astronauts to the Moon. There were 12 flights in total. The first two were unmanned to test out the new heat shields, and the following 10 were successful crewed flights.

Instead of the astronauts officially naming the capsule, these were all given the name Gemini followed by the mission number.

GEMINI 3 (*Molly Brown*)

Astronauts Gus Grissom and John Young were the first to fly a Gemini crewed mission and the first to perform the first orbital maneuver made by any crewed spacecraft. After Grissom jokingly named the capsule *Molly Brown*, after the Broadway musical *The Unsinkable Molly Brown*, it became the last and only Gemini flight NASA allowed the astronauts to name. The *Molly Brown* nickname was his way of not repeating what had happened to *Liberty Bell 7*, which sank after splashdown.

During this flight, John Young famously smuggled a corn beef sandwich onboard in the pocket of his spacesuit. While both astronauts did get a bite of the sandwich in space, the floating crumbs could have caused major issues and the sandwich was quickly returned to Young's pocket. Bread in its normal state hasn't been allowed onboard a NASA spacecraft since.

GEMINI 4

During the Gemini 4 flight, astronaut Ed White became the first American to leave the safety of the spacecraft and perform an EVA, or spacewalk. This was a key step in becoming ready to get to the Moon and to walk on the Moon.

GEMINI 5

This mission set a new flight record by staying in orbit for 8 days. It also tested out fuel cells that were used to generate electricity.

GEMINI 6A & 7

These missions were in orbit at the same time and performed the first space rendezvous. Gemini 7 set a new record for staying in space for 14 days.

GEMINI 8

On Gemini 8, the astronauts were able to dock with an unmanned vehicle while in space.

GEMINI 9A

Astronaut Eugene Cernan performed an EVA that was almost disastrous. There were no foot or handholds outside the spacecraft. So in a very stiff suit with no gravity, even simple tasks like turning valves was extremely difficult, because Cernan was floating with no way to create leverage. The suit was cooled by air, but became overwhelmed with the sweat and body heat of the astronaut, causing the helmet visor to fog up. At one point, Cernan had to use his nose to clear the fog from his visor to see. His spacewalk lasted only 2 hours and 7 minutes, but he was exhausted by the time he made it back to the airlock. At one point his heartrate reached 180 beats per minute.

Several important improvements were made after his spacewalk. Foot and handholds were added to the outside of the spacecraft and a cooling garment that used cold water that ran through thin tubes was worn next to the skin to control body temperature. These are both still used today.

GEMINI 10

On this mission, not only were they able to dock with another space vehicle, but astronauts were able to move both crafts using its engine.

GEMINI 11

Gemini 11 set a crewed Earth's orbit altitude record at 739.2 nautical miles (1,369 km). This record still stands as of 2021.

GEMINI 12

This was the last Gemini flight. Astronaut Buzz Aldrin performed a spacewalk and proved that meaningful work could be accomplished in space with the advantage of the new cooling garment, footholds and handholds and the ability to take breaks.[9]

 DID YOU KNOW?

The Gemini program was named after the Gemini constellation which is pronounced, "jeh-min-eye." NASA decided that their official pronunciation for Gemini would be different. NASA officials say "jim-in-ee," as in "Jiminy Cricket."[10]

The Apollo Missions (1961-1972)

The Apollo program was America's third space program, and the one that led to humans walking on the Moon.[11] In total, 24 men flew to the Moon and back, and 12 men walked on the Moon during the Apollo missions.

THE NAME APOLLO

To keep with the new tradition of using mythological names, Apollo was used for the mission to send humans to the Moon. In Greek mythology, Apollo is the god of the Sun. The man who named Project Mercury was Dr. Abe Silverstein, the former NASA Director of Space Flight Programs. He also came up with the name Apollo. In 1960, Silverstein was reading a book of mythology at home one night. He said the image of "Apollo riding his chariot across the sun was appropriate to the grand scale of the proposed program."[12]

TO THE MOON AND BACK

To fly to the Moon, land on the Moon and then fly back to Earth, engineers needed to design a new kind of spacecraft. What they came up with was the Command and Service Module (CSM) and the Lunar

Module (LM). The CSM was the main vehicle that would fly them to the Moon, orbit the Moon and then fly them home again. The LM was the 2-piece vehicle that would undock from the CSM, fly to the Moon's surface and land.

After the astronauts were done exploring the Moon, the LM would lift off from the Moon and rejoin the CSM. After the astronauts moved themselves and their equipment and souvenirs into the CSM, the CSM would release the LM to crash on the Moon so they wouldn't have to carry the extra weight back to Earth. Then the CSM fired its engines to make the return trip home.

COMMAND AND SERVICE MODULE

The CSM was the "mother ship" that carried three astronauts and the lunar module to the Moon. A total of 19 CSM were launched into space, with nine of them flying humans to the Moon and back. After the Apollo program, three CSMs were used to carry astronauts to the Skylab space station.

THE LUNAR MODULE

The lunar lander used by the Apollo program was originally known as the Lunar Excursion Model (LEM), and pronounced "Lem." It was later shortened to Lunar Module (LM) but still pronounced "Lem." The LM was attached to the CSM and was only used to transport astronauts from the Moon's orbit to the surface of the Moon. It had two parts: a descent stage and an ascent stage. The descent stage is what allowed the astronauts to land on the Moon and was left behind on the Moon. The ascent stage detached from the descent stage so that the astronauts could lift off from the Moon's surface and rejoin the orbiting CSM.

The LEM was a different-looking type of spacecraft with legs and the crew of Apollo 9 named their LM, "Spider" after its spider-like appearance.

To reduce weight, the LM had no seats, so the astronauts piloted the LM while standing/floating. The astronauts on the Moon slept in hammocks that could easily be put away while not in use.

APOLLO 1

The Apollo 1 ended in disaster when a fire broke out inside the capsule during a launch pad test on January 26, 1967. All three astronauts onboard were killed: Gus Grissom, Ed White and Roger B. Chaffee. Major changes were made as a result of the accident, including removing flammable material from the spacesuit and spacecraft cabin. Instead of using 100% pure oxygen—which allowed the fire to spread quickly—future missions used a mix of nitrogen and oxygen inside the cabin.

Apollo 1 was originally designated internally by NASA as "AS-204." The "A" stood for an Apollo spacecraft. The "S" meant it was using a Saturn launch vehicle. The "204" meant that it was the second version of the Saturn rocket and this was the fourth flight. At the request of the crew's families, it was renamed Apollo 1 posthumously to honor the astronauts that died and to recognize their mission properly among the Apollo missions.

UNCREWED TEST FLIGHTS

Apollo 4, 5 and 6 were uncrewed test flights to make sure the Apollo equipment would work when astronauts finally made it to the Moon.

CREWED TEST FLIGHTS

The first four crewed Apollo missions were also test flights. Apollo 7 was the first, and the crew tested out the Command Service Module in Earth's orbit.

APOLLO 7: FIRST APOLLO MISSION TO GET TO SPACE

Apollo 7 was the first American mission after the horrific Apollo 1 fire that killed all three men onboard. Following the fire, over 1,300 changes were made to the spacecraft. A few of these changes included redesigning the hatch allowing the crew to open it from inside in seven seconds and the outside crew to open it in ten seconds. All flammable materials were replaced with non-flammable materials, which included plastic switches being replaced with metal ones.[13]

Apollo 7 was the first to have a three-person American crew go into space and the first mission to do a live TV broadcast of Americans from space.[14] It was also the first mission where crew members caught colds which contributed to them refusing to accomplish tasks because they were overworked and sick.

APOLLO 8: FIRST MANNED MISSION TO THE MOON
Apollo 8 was the first manned mission to leave Earth's orbit and orbit the Moon. This was the first crewed flight to use the massive, 36-story tall, Saturn V rocket and the first to launch from Kennedy Space Center in Florida. The astronauts circled the Moon on Christmas Eve in 1968 and took the famous "Earthrise" photo.

APOLLO 9: SPIDER & GUMDROP
Starting with Apollo 9, the crews started using both the Lunar Module (LM) and the Command and Service Module (CSM). During this mission and each following Apollo mission, the crews gave them unique names. Apollo 9 was the first to test the Lunar Module, which they named Spider, due to its spider-like appearance. They gave their CSM the name Gumdrop. On Earth, the CSM was covered with blue wrappings that made it look like a wrapped gumdrop.[15]

APOLLO 10: SNOOPY & CHARLIE BROWN
This was the dress-rehearsal and last test mission before landing on the Moon. The crew tested all of the procedures and components needed for a lunar landing. They did everything except land on the Moon.

The Apollo 10 crew named their CSM Charlie Brown and the LM Snoopy after the Peanuts comic strip characters. The name Charlie Brown would be the guardian of the LM, Snoopy. The name Snoopy was given because it would be "snooping" around the Moon's surface.[16]

APOLLO 11: EAGLE & COLUMBIA
On July 20, 1969, the crew of Apollo 11 landed safely on the surface of the Moon. Over 6 hours later, Commander Neil Armstrong took his first steps on the Moon on July 21, 1969. Lunar module pilot Buzz

Aldrin joined him on the lunar surface about 19 minutes later. They spent over two hours on the first moonwalk, where they gathered 47.5 lbs (21.5 kg) of moon rocks and other lunar samples.

The third Apollo 11 crew member, Michael Collins, flew the CSM Columbia alone around the Moon for over 21 hours before Armstrong and Aldrin rejoined him.

THE EAGLE HAS LANDED

"Houston, Tranquility Base here. The Eagle has landed." These were Neil Armstrong's words broadcast from the surface of the Moon to officially announce that they had safely landed on the Moon. The name of the Apollo 11 Lunar Module was Eagle and they had landed on a site called Tranquility Base in a relatively smooth and flat area on the Moon called, the Sea of Tranquility. It was named Eagle after the bald eagle on the mission patch.

The Command Module was named Columbia. It was named after Columbiad, the canon used in Jules Verne's 1865 novel, *From the Earth to the Moon.*[17]

 DID YOU KNOW?
Eagle and Columbia were not the original names for the command module and lunar module. The Apollo 11 crew had originally named the command module Snowcone and the lunar module Haystack after their shapes. However, NASA felt like these names weren't very inspiring, so the crew was encouraged to try again.[18]

APOLLO 12: YANKEE CLIPPER & INTREPID

The November after Apollo 11 landed on the Moon, NASA sent three more men back to the Moon on Apollo 12. Apollo 12 was struck by lightning twice, first at 32 seconds and then again at 52 seconds after it launched.[19] While it caused some instrument problems, there was little damage and they successfully landed in an area of the Moon called, the Ocean of Storms, which is the largest dark spot on the Moon. The Apollo 12 crew performed two EVAs or moonwalks, on the Moon and were able to collect 73.75 lbs (33.45 kg) of moon rocks and samples.

Yankee Clipper, the name for the CSM, was selected from names submitted by employees who had built it. The LM was named Intrepid, which also came from names submitted by those who had built it.

"WHOOPIE!"

The first words spoken by Apollo 12 Commander Pete Conrad, the third man to step on the Moon, were "Whoopie! Man, that may have been a small one for Neil, but that's a long one for me."[20] At 5 foot, 6 inches (168 cm) tall, he was one of the smaller Apollo astronauts.[21] For comparison, Neil Armstrong was 5 feet, 11 inches (180 cm) tall.

APOLLO 13: ODYSSEY & AQUARIUS

Apollo 13 didn't land on the Moon as intended. When an oxygen tank exploded two days into the mission, the plan to land on the Moon was aborted and the only mission that remained was to safely get the crew back to Earth.

The command and service module was damaged, so the crew moved to the lunar module, which then acted as a lifeboat. As they looped around the Moon, the astronauts and crews on the ground worked together to solve many problems. The first was to figure out a way to use the LM—which was built to support two astronauts for 45 hours on the Moon—to support three astronauts for the three or four-day journey back to Earth. During their journey, they set the human record for traveling the furthest distance from Earth.[22]

Astronauts Jim Lovell, Jack Swigert and Fred Haise all miraculously made it safely back to Earth. They all were in fairly good condition, except for Fred Haise. Because of a lack of water, a urinary tract infection that he was suffering from turned into a kidney infection and he was in pain for most of the journey.

The crew of Apollo 13 named their CSM Odyssey. It means a long voyage marked by changes of fortune. It reminded the crew of Homer's ancient Greek poem, Odyssey and the long voyage made by Odysseus.[23]

They named their lunar module Aquarius after the Egyptian god of the same name, which means "water carrier." Aquarius brought life and knowledge to the Nile Valley, and the crew of Apollo 13 hoped to bring back knowledge from their trip to the Moon.

 DID YOU KNOW?

HOUSTON, WE'VE HAD A PROBLEM WAS SAID TWICE BY TWO DIFFERENT PEOPLE

While Jim Lovell's recording is often the one heard saying, "Houston, we've had a problem," astronaut Jack Swigert actually said it first. He said, "OK Houston, we've had a problem here." Mission Control responded with, "This is Houston. Say again, please." Jim Lovell then said, "Uh, Houston, we've had a problem." On the official recording, Swigert's voice is garbled at the beginning, while Jim's recording is clear, which is why it's the one most often played.[24]

APOLLO 14: KITTY HAWK & ANTARES

The crew of Apollo 14 successfully landed on the Moon's lunar highlands where Apollo 13 had planned on landing. The crew of Apollo 14 included Alan Shepard, Stuart Roosa and Edgar Mitchell.

Alan Shepard, the first American in space, made it back to space a second time and became the fifth person to walk on the Moon. At the age of 47, this also made him the oldest person to walk on the Moon. He also famously hit two golf balls on the surface of the Moon with a makeshift golf club.[25]

Apollo 14 gathered a total of 94 lbs (43 kg) of moon rocks and samples. One of these was named Big Bertha, a 19.8 pound (8.9 kg) moon rock. Scientists learned that Big Bertha contains an embedded meteorite from Earth that is 4 billion years old, which makes it the oldest known Earth rock.[26]

Kitty Hawk is the name of Apollo 14's CSM. It is named after both the location of the Wright brothers' first flight in North Carolina and the name of their first successful powered aircraft, which was also called the Kitty Hawk Flyer, or Wright Flyer. Their LM is named after the star Antares that was used to orient the module when landing on the Moon.[27]

MOON TREES

While Shepard and Mitchell were on the Moon for two days, Command Module Pilot Stuart Roosa orbited the Moon alone 34 times onboard *Kitty Hawk*. He took with him 500 tree seeds from five different species: redwood, Douglas fir, sweetgum, sycamore and loblolly pine.

After landing, Roosa sent them to the U.S. Forest Service. Nearly all of the seeds germinated successfully and between 420 to 450 seedlings were then planted across the U.S. and became known as Moon Trees. Many were planted next to their counterpart on Earth as an experiment to see if there were any differences as they grew. Many were given away to forest services in other states. One of the Loblolly Pine was planted at the White House.[28]

APOLLO 15: ENDEAVOR, FALCON & THE MOON BUGGY

Apollo 15 was the fourth mission to land on the Moon and with it, The Lunar Rover, also known as The Moon Buggy. The Moon Buggy made it possible for the crew to travel farther from the LM than ever before. This particular Moon Buggy came with a plaque that read, "Man's First Wheels on the Moon, Delivered by Falcon, July 30, 1971."[29]

Apollo 15's crew included David Scott, James Irwin and Alfred Worden. Highlights from this mission included the first and only standup EVA or spacewalk from the hatch on top of the Lunar Module. This allowed astronaut David Scott to check out where they had landed and plan out how best to perform the next day's EVA.[30] They collected about 170 lbs (77 kg) of lunar rocks and samples, including the Genesis Rock, which is believed to be part of early lunar crust. On the way back to Earth, the first deep space EVA was performed by Alfred Worden.

Endeavour was the name of Apollo 15's CSM and was named after the ship used by Captain James Cook's scientific voyages in the 18th century. Falcon was the name of the lunar module and was named after the United States Air Force mascot. All three astronauts onboard Apollo 15 were in the Air Force. The feather used in this mission's famous hammer and feather experiment (see next fact) was from an Air Force Academy's female falcon mascot.[31]

THE FEATHER & THE HAMMER, A BIBLE & THE FALLEN ASTRONAUT

Scott did several other things that made for a memorable trip. In front of the television camera, he took a hammer and a feather and dropped them at the same time. According to Galileo's theory, all objects fall at the same rate, regardless of mass, when there is no drag. The hammer and feather hit the ground at the same time.[32,33]

Away from the camera, Scott also drove a short distance away from the LM and left behind a small aluminum figure that was meant to represent an astronaut in a spacesuit called The Fallen Astronaut. With it, he placed a plaque with the names of 14 known astronauts and cosmonauts who had died as part of space exploration.[34] Before entering the LM to leave the Moon, Scott left a small red Bible on the control panel of the Moon buggy.[35]

APOLLO 16: CASPER & ORION

Astronauts Charlie Duke, John Young and Ken Mattingly made up the crew of Apollo 16. Duke and Young spent just under three days on the Moon and performed three moonwalks. This was Young's fourth spaceflight. Mattingly stayed on the CSM and orbited the Moon 64 times. Duke was originally assigned to Apollo 13, but had been grounded after being exposed to the rubella virus. Even though Duke had been selected as an astronaut in 1966, and had been CAPCOM for Apollo 11, this was his first and only flight to space. At the age of 36, he holds the record for being the youngest person to walk on the Moon.

The Moon Buggy was again part of the mission and was driven over 16 miles (26.7 km). It was also used to gather the biggest moon rock brought back to Earth, called Big Muley. It weighed 26 lbs

(11.7 kg). In total, this mission brought back 211 lbs (95.8 kg) of moon rocks.

Duke left a photo of his family on the surface of the Moon and named a crater CAT after his sons **C**harlie **a**nd **T**om and another crater, Dot, after his wife, Dorothy.[36]

Because the white Teflon spacesuits worn by astronauts on the Moon looked shapeless on TV, the Apollo 16 CSM was named Casper after the cartoon character, Casper the Friendly Ghost. The LM was named after the constellation Orion. It was one of the stars they used to navigate with while in space.[37]

APOLLO 17: AMERICA & CHALLENGER

Apollo 17 was the sixth and final mission to the Moon. Eugene Cernan, Harrison "Jack" Schmidt and CSM pilot Ronald Evans were the last humans to fly to the Moon and back.

They broke several records during their mission, including bringing the largest collection of moon rocks (243.7 lbs or 110.52 kg), spending the longest time in orbit (6 days and 4 hours) and having the most lunar orbits (75).

During one of their three EVAs, they made the exciting discovery of orange soil, which turned out to be small beads of volcanic glass that were over 3.5 billion years old.

As a thank you and a tribute to the American people who made the trip to the Moon possible, the final CSM was named America. The lunar module was named Challenger to represent the challenges that lie ahead, for Apollo and beyond.[38]

THE BLUE MARBLE

Probably the most meaningful thing they brought back from their mission was a color picture of an entirely illuminated, whole, round Earth. It's called the Blue Marble because it resembles a round, glass marble. Because of landing sites and the timings of the Moon and Sun, the Apollo 17 crew were the only crew to have seen the Earth in its entirety without a shadow. This is one of the most reproduced images in history.[39]

The Blue Marble from Apollo 17. Image courtesy of NASA Johnson Space Center Gateway to Astronaut Photography of Earth.[40]

MOON EXPLORATION COMMORATIVE PLAQUE

A final Apollo plaque on Challenger's strut was unveiled to commemorate the completion of man's first exploration of the Moon.[41] Commander Eugene Cernan made these final remarks:

> "This is our commemoration that will be here until someone like us, until some of you who are out there, who are the promise of the future, come back to read it again and to further the exploration and the meaning of Apollo." [42]

LAST WORDS ON THE SURFACE OF THE MOON

We all know the first words spoken as the first man stepped on the Moon, but very few know the last words spoken on the surface of the Moon. Commander Eugene Cernan was the very last man on the Moon and he said these words right before he and Schmitt packed up their samples, discarded the tools they no longer needed and boarded the Challenger LM, "*I believe history will record that America's challenge of today has forged man's destiny of tomorrow.*" [43]

QUARANTINE FROM MOON GERMS

Just in case the returning astronauts brought back a case of the Moon plague, crews from Apollo 11, 12 and 14 spent three weeks in quarantine when they returned home. For the first few days they actually stayed in a converted Airstream trailer known as the Mobile Quarantine Facility, or MQF.

Four MQFs were built for Apollo 11 through 14, but only three of the crews used them. This unit was mobile, which allowed it to be transferred on sea, air and road all while keeping the astronauts separate from everyone. After Apollo 14, NASA decided there were no dangerous moon germs Earth needed protecting from and did away with quarantining after missions. [44]

The Space Shuttle Program (1981- 2011)

The Space Shuttle program lasted from 1981 to 2011. During this time, five different space shuttles flew a combined total of 135 missions with 355 people—representing 16 countries—onboard. The shuttles flew to the International Space Station 37 times.

HOW MANY SHUTTLES?

Six space shuttles were built, but only five were operational. The first, *Enterprise*, was built in 1976 and was only used in training for approach and landing tests. The other five were fully functional and space-rated. They were named: *Columbia, Challenger, Discovery, Atlantis* and *Endeavour*. Each space shuttle could hold up to eight people.

DID YOU KNOW?

The famous picture of Earth we know as "The Blue Marble," was actually taken upside down with the South Pole appearing at the top of the globe, not at the bottom.[45] This was because it

was taken by a man who was weightless, upside down at the time and just wanted to get this amazing shot. Most of the reproductions we've seen of the photo have been cropped and inverted from the original picture, officially named AS17-148-22727, to match our expectations.[46]

ENTERPRISE (OV-101)

Enterprise was the first space shuttle, or Orbiter Vehicle, to be built. It was originally going to be named *Constitution*, and debuted on Constitution Day (September 17, 1976) in honor of the U.S. Constitution's Bicentennial. But *Star Trek* fans had hoped they could persuade the U.S. president to name it after the show's famous starship, USS *Enterprise*. Hundreds of thousands of letters later and President Ford asked NASA officials to change the name to *Enterprise*.[47]

The fictional USS *Enterprise* of *Star Trek* is on a mission to "explore strange, new worlds; to seek out new life and new civilizations; to boldly go where no man has gone before." With a mission statement like that, who could blame them for wanting to change the name of such a historic spacecraft?

In response to the name change, *Star Trek* creator Gene Roddenberry and many of the original *Star Trek* cast were there to witness *Enterprise* being unveiled during the rollout ceremony on Constitution Day September 17, 1976.

COLUMBIA – FLEW 28 MISSIONS

Space Shuttle *Columbia* was the first space shuttle to fly into space. It gets its name from a few different places. One is after the first American vessel to circumnavigate the globe, the *Columbia Rediviva*. It's also named after the Apollo 11 command module that brought the first men who walked on the Moon. Columbia is also the name of the female symbol or personification of the United States and comes from Christopher Columbus.[48]

Columbia was the second shuttle destroyed in flight, killing everyone onboard on February 1, 2003, at the end of its 28th mission.[49]

CHALLENGER – FLEW 10 MISSIONS

Challenger was the second space-rated orbiter to fly into space. She carried the first American woman—Sally Ride—into space and was also the first space shuttle to have been involved in an accident that led to the death of the entire crew.

It was named after Apollo 17's Lunar Module, which bore the same name. Its name also comes from the ship HMS *Challenger*, which made groundbreaking discoveries that laid the foundation of oceanography that we know today.[50]

DISCOVERY – FLEW 39 MISSIONS

Discovery flew 39 missions over 27 years, which was the most flights of any space shuttle. To stay with naming shuttles after exploration ships, *Discovery* was named after several famous ships: Captain James Cook's HMS *Discovery*, which visited the Hawaiian Islands, the RRS *Discovery* that went to Antarctica and another HMS *Discovery* that made an expedition to the North Pole.[51]

ATLANTIS – FLEW 33 MISSIONS

Atlantis was named after the famous RV *Atlantis* that served as the main research vessel for the Woods Hole Oceanographic Institution from 1930 to 1966.[52] *Atlantis* also made an appearance in a few movies, including *SpaceCamp*, *Deep Impact* and *Armageddon*. *Atlantis* made the final flight that marked the end of the Space Shuttle program on July 21, 2011.

ENDEAVOUR – FLEW 25 MISSIONS

Endeavour was the last space shuttle to be built, and was to replace *Challenger* after it was destroyed. *Endeavour* was built using spare parts from *Discovery* and *Atlantis*.[53]

In a nationwide contest to name the newest space shuttle, school children voted to name it *Endeavour* after the 18th century British ship HMS *Endeavour* captained by James Cook. This is why it has the British English spelling. Apollo 15's command module was also named *Endeavour*.

To honor its namesake, the *Endeavour* carried a piece of wood from the original *Endeavour* ship inside the cockpit.[54]

FIRST SHUTTLE LAUNCH

April 12th is a special day. It marks the day in 1961 that Soviet cosmonaut Yuri Gagarin became the first human in space. Exactly 20 years later, in 1981, NASA launched the first reusable space plane, the Space Shuttle *Columbia*. STS-1 successfully launched a two-man crew into space, orbited the Earth 37 times and glided back safely.

THE LONGEST SPACE SHUTTLE MISSION

Space Shuttle *Columbia* set the record for the longest space shuttle mission in 1996 with a 17-day mission.[55]

A MASSIVE GLIDER

The Space Shuttle does not use its engines upon re-entry, which basically means it's a massive glider and has only one shot at landing properly and safely. There was no way to try again if the shuttle wasn't lined up properly or was coming in too fast.

DIFFICULT TO FLY

The space shuttle was one of the hardest planes to fly and to land. In the cockpit, there were more than 2,000 switches and circuit breakers. On the entire space shuttle, there were more than 2 1/2 million moving parts [56] and hundreds of ways to do something wrong.[57]

DROPPING THE WHEELS

Dropping its landing gear made flying the space shuttle even more difficult. Because of this, the pilot would wait as long as possible to drop the landing gear. So rather than drop the wheels at 1,800 feet (549 meters) above ground level like most commercial airplanes, the space shuttle would drop its wheels at 1,000 feet (305 meters) above ground level.

LONG RUNWAYS

The space shuttles could only land on the longest runways on Earth. There were three landing sites in the U.S. The main one was at the Kennedy Space Center in Florida, where 78 space shuttles landed. The second one was in California at the Edwards Air Force Base, where 54 space shuttle missions landed. New Mexico had an emergency shuttle landing site at the White Sands Space Harbor. STS-3 *Columbia*, was the only space shuttle mission to land there due to flooding in the originally planned landing site in California in 1982.[58] Other international shuttle landing sites could be used in emergencies, including ones in Spain, France, Germany, Morocco and Easter Island.

TOOK ABOUT 2 DAYS TO DOCK WITH THE ISS

While the Soyuz can dock with the ISS in about 6 hours,[59] the Space Shuttle took much longer. It took the space shuttles about 50 hours, or 2 days, to dock with the International Space Station.[60]

SPACE SHUTTLE DISASTERS

Fourteen astronauts were killed during the Space Shuttle program on two different space shuttles, *Challenger* in 1986 and *Columbia* in 2003. Compared to other space programs, the space shuttle program had the most fatalities.

COST TO LAUNCH

Over the course of the space shuttle program, NASA and the U.S. Congress spent more than $192 billion. With 135 launches, this works out to be about an average of $1.4 billion per launch. Some of the shuttle equipment used for each launch wasn't reusable, which kept the costs high.

With new spacecraft, costs per launch are much less due to more reusable parts. For example, the cost per launch for the SpaceX's Falcon 9 is $62 million for the first launch and $50 million for subsequent launches.

RUSSIA'S SPACE SHUTTLE

Buran means "snowstorm" or "blizzard" in Russian. The Buran was Russia's spaceplane shuttle orbiter, but unlike the U.S. space shuttles, the Buran could be flown on autopilot.[61] The Buran only made one successful uncrewed test flight. While there were plans for future flights, the program never "got off the ground" and was suspended due to budget cuts and for political reasons. Two more orbiters were planned, but never completed. One of them is known unofficially as Ptichka which means "storm" or "tempest."

Soviet Space Programs

THE VOSTOK PROGRAM (1959-1963)

Vostok means "east" or "orient" in Russian. This was the Soviet Union's first space program. From 102 candidates, the Soviet Union selected 20 air force pilots to train as cosmonauts in early 1960. Six of those 20 were selected for more advanced training and became known as the "The Vanguard Six" or "The Sochi Six." The six men who were selected were the shortest of the original 20.[62] Because of how small the Vostok capsule was, applicants had to be 5 feet, 7 inches (170 cm) or shorter and weigh 154 lbs (70 kg) or less.

The Vostok spacecraft were more automated than the American equivalent, so less emphasis was put on piloting experience. The Soviets looked at men who had been exposed to higher g-forces, worked well in high-stress situations and were familiar with using ejection seats.

VOSTOK 1: YURI GAGARIN

Cosmonaut Yuri Gagarin was one of "The Vanguard Six," and was selected to be the first to fly onboard Vostok 1. At the time, the Soviets' success rate was 50%—12 out of 24 were successful launches. There were several nervous people in ground control, but Gagarin was reported to have been calm while waiting to launch, with a super low heart rate of 64 beats per minute. On April 12, 1961, Gagarin let out a "Poyekhali! ("Let's Roll!") at liftoff as he launched and became the first human in space. His flight included one orbit around Earth. During re-entry he experienced over 8 g's or 8 times the force of Earth's gravity (people can experience 4 to 5g's briefly on a rollercoaster, but they are designed specifically to be for short periods so you don't pass out). Gagarin successfully ejected from the capsule, and both he and the capsule parachuted safely to the ground. He returned to Earth a hero.

COSMONAUTICS DAY

April 12th was declared Cosmonautics Day after Gagarin's successful launch and landing and is one of Russia's official holidays.

VOSTOK 2: GHERMAN TITOV

On August 6, 1961, Cosmonaut Gherman Titov was just under 26 years old when he became the second human in space. He remained the youngest person to fly into space until July 20, 2021 when 18-year-old Oliver Daemen launched into space onboard Blue Origin's first crewed New Shepard flight. Titov launched onboard the Vostok 2 and spent an entire day in space, orbiting the Earth 17.5 times, which set a record Americans didn't reach until two years later. He experienced many firsts, including the first sleep in space and the first space sickness.

VOSTOK 3 & 4: ANDRIYAN NIKOLAYEV & PAVEL POPOVICHALSO

Vostok 3 launched on August 11, and Vostok 4 followed by launching the next day, August 12, 1962. They became the first simultaneous spaceflights. Onboard Vostok 3, Cosmonaut Andriyan Nikolayev unstrapped himself and became the first human to free float in microgravity. Both vehicles were in sight of each other for much of their flights and they were able to communicate with one another via shortwave radio.

VOSTOK 5: VALERY BYKOVSKY

Cosmonaut Bykovsky was set to spend eight days in space, but solar flare activity caused his mission to be shortened to five days instead. However, he still set a record for the longest solo space flight, which remains today.

VOSTOK 6: VALENTINA TERESHKOVA

The Soviet Union again set another first by sending the first woman into space. Valentina Tereshkova spent almost three days in space, orbited earth 48 times and performed several experiments to test how female bodies functioned in space. Like her male cosmonaut counterparts, she kept a flight log, took pictures and manually made adjustments to the spacecraft's orientation. One of her photos of the Earth's horizon was used to identify aerosol layers in Earth's atmosphere. Tereshkova remains the only woman to have been on a solo space flight mission.

The Vostok capsule was very small and to spend almost three days in such cramped conditions in a spacesuit was hard. Tereshkova experienced nausea and physical discomfort for most of her flight.[63] But one of her biggest complaints during the flight was that while the Soviet team had packed her capsule with food, water and toothpaste, they forgot to include a toothbrush! [64]

COSMONAUT CALLSIGNS

Rather than naming individual space capsules like the Mercury capsule and the Apollo spacecraft, the cosmonauts have individual callsigns, starting with the first man in space being called Cedar. The tradition continues today, and Cosmonaut Ryzhikov has the callsign Favorites.[65]

Here are the callsigns for the six cosmonauts who flew on Vostok missions:

CEDAR: Yuri Gagarin was known as Kedr which means "cedar" in Russian. It was a common sounding name which was done on purpose. The Soviets were very secretive about their space program and didn't want anyone who might have overheard radio conversation to know who or what was being referred to.[66]

EAGLE: Gherman Titov's callsign was Oryol, which translates to "eagle."

FALCON: Andriyan Nikolayev flew on two spaceflights and had the callsign Falcon for both missions.

GOLDEN EAGLE: Pavel Popovich also flew twice and his callsign was Golden Eagle.

HAWK: Valery Fyodorovich Bykovsky set the record for longest solo space flight and was known as Hawk.

SEAGULL: The first woman in space, Valentina Tereshkova, was Chayka or "seagull."

Voskhod Program (1964-1966)

Following the Vostok program, the next Soviet space program was the Voskhod program. It was similar to America's Gemini space program. Voskhod means "ascent" or "dawn" in Russian, and was perhaps used to represent the dawn of a new space age. The Voskhod spacecraft was basically a Vostok spacecraft, but slightly larger. This allowed it to accommodate two cosmonauts rather than one. One major difference between the two spacecrafts was that the Voskhod capsule landed with the crew inside rather than the cosmonaut ejecting and parachuting to the ground separately from the capsule.

The main objective of the Voskhod program was to launch more than one person into space and to better understand the effects of space travel and microgravity on the human body.

VOSKHOD 1 – FIRST MULTIPERSON CREW IN SPACE

Again, the Soviets beat the American in the space race by sending more than one human into space on one spacecraft onboard Voskhod 1 in October 1964. The Voskhod was built to fit two cosmonauts, but three cosmonauts squeezed into the small Voskhod capsule and made history. It was a tight fit with three men inside and they went without spacesuits, making this flight the first without spacesuits. The flight only lasted one day.

VOSKHOD 2 – FIRST SPACEWALK & HEROIC LANDING

On the second Voskhod flight in March 1965, only two men were onboard and both had spacesuits because the main objective for the one-day flight was to perform the first ever spacewalk. Cosmonaut Alexei Leonov spent 12 minutes outside the Voskhod spacecraft and became the first person to walk in space.

Cosmonaut Pavel Belyayev had to take control of the spacecraft and fly it back to Earth manually after the automatic re-entry systems failed. They safely landed 240 miles (386 km) off course in a forest in the Ural Mountains in deep snow and spent a chilly night inside their capsule waiting for help. Due to the thick forest that surrounded them, it took two days before they were rescued.

KOSMOS 110 – 2 DOGS ORBIT EARTH FOR 22 DAYS

As part of the Voskhod program, 2 dogs named Veterok ("Light Breeze") and Ugolyok ("Ember") launched into orbit on Feb 22, 1966 and returned safely 22 days later on March 16, 1966. This set the record that still stands as the longest space flight by dogs.[67]

The Luna Program (1959-1976)

The Luna Program was the Soviet Union's space program to gather information about the Moon, including photographs, measuring the Moon's gravity, temperature, chemical composition and radiation levels. The program's first spacecraft, Luna 1, launched in January 1959. The last Luna spacecraft, Luna 24, returned to Earth with samples from the Moon in August 1976.[68]

The Luna program achieved several Lunar firsts, including: the first probe to impact the Moon, first flyby and image of the far side of the Moon, first soft landing, first lunar satellite, the first analysis of lunar soil, the first sample return and the first lunar rover deployment.[69]

The Luna Program included several types of spacecrafts including:

Impactors

Impactors were designed to hit the near side of the Moon and send photos back to Earth before it crashed into the surface. The U.S.'s version of the Luna's impactors was their Ranger Program.

Flybys

Flybys were simple spacecrafts meant to fly around the Moon and send photos back to Earth. Luna 3 sent the first photos of the far side of the Moon, which is the side of the Moon that can never been seen from Earth. The competing U.S. version was called the Pioneer program.

Soft Landers

Soft Landers were exactly what they sound like, spacecrafts designed to land softly on the Moon. Their purpose was to not only take photos of the Moon, but also to take samples of the Moon's surface and collect other information. The first close-up shots of the Lunar surface were taken by the Luna 9 soft lander. The U.S. sent a few soft landers of their own with their Surveyor program.

Orbiters

Orbiters did just that, they were sent to orbit the Moon. Luna 10 became the Moon's first satellite. The U.S. sent seven orbiter probes as part of their Pioneer program, but they all failed. It wasn't until their Lunar Orbiter program finally succeeded and was able to map out 99% of the Moon's surface that it was possible to select the landing sites for the Apollo program.

Rovers

The Luna program was able to deploy two lunar landers on the surface of the Moon. Luna 17 and Luna 21 carried Lunokhod lunar rovers. The first lunar rover on the Moon landed in 1970 and was able to send back more than 20,000 television images and 205 panoramic high-resolution images.[70] The first U.S. lunar rover arrived with the crew of Apollo 15 in 1971.

Sample Return

A Sample Return spacecraft was an advanced soft lander that could scoop up lunar samples and then launch back to Earth with those samples. The Soviet Union had three successful Sample Return Luna missions: Luna 16 in 1970, Luna 20 in 1974 and Luna 24 in 1976. In total, they returned 10.6 ounces (301 grams) of lunar soil samples back to Earth.

The Americans' approach was to send men to the Moon and return samples. They were successful six out of seven times and were able to return 839.87 lbs (380.96) kg of moon rocks and soil samples.

LUNA 15 & APOLLO 11

When the crew of Apollo 11 reached the Moon's orbit in July 1969, the Soviet Union's Luna 15 unmanned Sample Return spacecraft was already in orbit. Its purpose was to retrieve moon soil samples, and if it happened to beat the manned Apollo 11 crew back, even better.

Luna 15 had difficulty finding a suitable landing spot and was delayed in landing before Neil Armstrong and Buzz Aldrin set foot on the Moon. Two hours after Aldrin and Armstrong lifted off the surface of the Moon after their historic landing, Luna 15 crashed-landed into a mountain on the Moon.[71]

SOVIETS PLANNED TO SEND HUMANS TO THE MOON

During the time when Americans were working frantically and spending billions to be the first to send a man to the Moon, the Soviets denied trying to do the same and claimed instead that they were focused on sending lunar satellites and sending robotic probes to the Moon.

However, it turns out that they had two crewed lunar programs they were secretly pursuing. One was to send men on a crewed lunar flyby and the second to land men on the Moon.[72] The Soviets built lunar landers, but had several N-1 rocket failures that were critical to their crewed lunar missions. After the Americans successfully sent human crews to the Moon, both Soviet lunar programs were eventually cancelled.[73]

The Soyuz Program (1966–Present)

Soyuz means "union" in Russian and was the third Russian space program.

THE ROCKET AND THE SPACECRAFT

Soyuz is the name of the rocket as well as the spacecraft. Soyuz rockets have flown more than 1,700 times, which makes it the world's most used space rocket.

IF IT AIN'T BROKE, DON'T CHANGE IT

Russia's Soyuz series of spacecraft is the longest-serving manned spacecraft in the world. More than 220 capsules have been built and over 400 people on 141 crewed missions have launched to space in them.

IMPRESSIVE SAFETY HISTORY

The Soyuz spacecraft is one of the most reliable and safest rockets to fly to space. The Soyuz has a launch escape system that allows a survivable return to earth anytime between the point from launch to orbit. When a fire began at the base of the rocket while still on the launch pad on September 26, 1983, this launch escape system saved two Russian cosmonauts' lives: Vladimir Titov and Gennady Strekalov.[74]

Russia's Soyuz program has only had four deaths: Cosmonaut Vladimir Komarov on Soyuz 1 in April 1967, and three cosmonauts, Georgy Dobrovolsky, Viktor Patsayev and Vladislav Volkov, on Soyuz 11 in June 1971.

THREE SEATS TO SPACE

The Soyuz has three seats, with the commander sitting in the middle seat. The left seat is for Flight Engineer 1, who is the backup commander. The right seat is for Flight Engineer 2. The Soyuz can only be flown manually from the middle seat using two hand-controllers. Normally, the Soyuz is not flown manually. That is only done in special circumstances.

CUSTOM-MADE SEATS

Each seat in the Soyuz is custom-made to fit the astronaut who sits in it. Plaster molds are taken of each astronauts' body and the seats are made from those molds. This isn't exactly for comfort, it's more about saving space and for safety. The custom-made seats cradle the spine and absorb some of the hard impact when the Soyuz returns to the Earth.

SOYUZ CARGO SPACE

On return to Earth, the Soyuz has space for about 110 lbs. (50 kg) of cargo in addition to the crew of three.

HIBERNATION MODE

While docked to the ISS in hibernation mode, the Soyuz can remain in space for 210 days.[75]

CAN NAVIGATE VISUALLY USING A PERISCOPE

The Russians have a simple, but effective, way of navigating onboard the Soyuz. It has a periscope that allows astronauts to look out of their ship, find Earth and align it with the periscope. This helps them know where they are and how to get to where they want to go.

ISS LIFEBOAT

The Soyuz acts like a lifeboat for the International Space Station. At least one Soyuz is always attached to the space station. In case of an emergency, the ISS crew could use the Soyuz to return to Earth.[76] The Soyuz capsules are centrally located, their command-and-control computers are always on, there are paper copies of the emergency procedures and a variety of radio panels that can be reconfigured to communicate with both Moscow and Houston if needed.[77]

ASSEMBLED HORIZONTALLY

One of the differences between NASA and Roscosmos space vehicles is in how they are assembled. Rather than assembled vertically and then moved to the launchpad balanced on the enormous crawler transporter, the Soyuz is assembled horizontally and rolled out to the launchpad in that position. It isn't until a couple of days before launch that the Soyuz is repositioned vertically and pointed toward space.

RAIN OR SHINE, THE SOYUZ CAN LAUNCH

NASA has a long history of launch delays due to weather related conditions like high winds, cloud cover and cold temperatures. But Roscosmos can launch the Soyuz in almost any weather condition.[78]

NO LIFE RAFT ON THE SOYUZ

Unlike many other space vehicles, the Soyuz was built to float and does not carry a life raft onboard. Due to where the Soyuz launches from and the mass amounts of land that Russia has, the Soyuz capsules almost always land on land and not on the water.[79] The only Soyuz to have needed an emergency water landing was Soyuz 23, which ended up in Kazakhstan's Lake Tengiz in 1976.[80]

NO FIRE EXTINGUISHER ON THE SOYUZ

Space is precious onboard a spacecraft and the crews onboard the

Soyuz are trained to handle the event of fire without the use of a fire extinguisher. In case of fire, the crew would close their helmets and depressurize the whole spacecraft by releasing all of the cabin's oxygen. Where there is no oxygen, there is no fire. Once the fire is out, they pressurize the cabin with the appropriate amount of oxygen and then open their helmets and continue their mission.[81]

ALWAYS TWO SEARCH AND RESCUE TEAMS
To be prepared, there are two search and rescue teams ready to meet up with the returning Soyuz. Since landings don't always occur where they are supposed to, there's a team in the planned landing site and one to be sent to a second site if it lands off course.[82]

NO MANUAL RE-ENTRY
While every commander and flight engineer onboard the Soyuz is trained to perform a successful re-entry, there has never been a manual re-entry in the history of the Soyuz. The Soyuz has always re-entered on autopilot.[83]

SPEEDY RENDEZVOUS
The Space Shuttle would often take 50 hours—just over two days—to dock with the Space Station. The Soyuz is able to perform what's known as a "quick rendezvous" and can dock with the space station six hours after launch, or in four orbits as the Soyuz crew would say.[84]

China's Shenzhou Space Program
Shenzhou, roughly translated from Chinese means, "divine vessel," "divine ship" or "divine craft." The Shenzhou Space program put the first Chinese taikonaut, Yang Liwei, into space. Russia has partnered with China and has helped China develop their space program and transferred much of their crewed spacecraft technology to China. This included Soyuz technology, space suits, life support systems and cosmonaut training.

SHENZHOU SPACECRAFT

Because their technology came from Russia, the Chinese Shenzhou spacecraft looks a lot like Russia's Soyuz, though there are differences. The Shenzhou is larger, and carries an inflatable life raft in case of a water landing, whereas the Soyuz doesn't. The seating arrangements are also slightly different. In the Shenzhou the commander sits in the center seat just as in the Soyuz, but the copilot sits in the left seat. In the Soyuz, the copilot sits to the right of the commander.[85]

SHENZHOU 5: FIRST TAIKONAUT IN SPACE

China's first manned spaceflight launched October 15, 2003. It made Yang Liwei the first Chinese citizen to fly into space. Liwei became the first Chinese astronaut, or taikonaut, and orbited the earth 14 times in just under one day.

SHENZHOU 6: FIRST MULTI-DAY SPACE MISSION

China's second crewed launch in 2005 sent two men into space for four days, making this their first multiple-day space mission.

SHENZHOU 7: FIRST THREE-PERSON CREW & SPACEWALK

In September of 2008, China sent three people into space at one time and performed their first spacewalk, which lasted 22 minutes. The taikonauts wore the Chinese-developed Feitian spacesuit.

SHENZHOU 9: FIRST CHINESE WOMAN IN SPACE & DOCKING TO SPACE STATION

In June of 2012, Liu Yang became the first female taikonaut. This mission had another first. The Shenzhou 9 crew docked with and boarded China's Tiangong-1 space station. They stayed onboard for six days and then returned safely home.

SHENZHOU 10: SECOND CHINESE WOMAN IN SPACE

Shenzhou 10 launched a year later in June 2013. With the second female taikonaut onboard, it became the last crew to board China's first space station, Tiangong-1, before it was retired. During their record breaking 15-day mission, female taikonaut Wang Yaping gave a video lecture to students across China while in orbit. The crew also performed some space station maintenance and conducted science experiments.

SHENZHOU 11: FIRST AND ONLY DOCKING TO TIANGONG-2 SPACE STATION

A month after China's second space station was launched into orbit in September 2016, the 2-man crew of Shenzhou 11 docked with and boarded Tiangong-2. Their 33-day mission set a new record to become the longest Chinese crewed space mission. One of the crew members, Jing Haipeng, turned 50 while in space. This was also his third spaceflight, which also makes him the taikonaut with the most flights.

 DID YOU KNOW?
While Yang Liwei was the first Chinese citizen to fly into space in 2003, he was not the first person of Chinese origin to fly into space. That honor goes to Taylor Wang, who was born in Shanghai, but became a U.S. citizen in 1975 and flew onboard the Space Shuttle in 1985.

SpaceX's Dragon

THE ORIGINAL DRAGON

Dragon 1 was built as a reusable spacecraft meant to deliver cargo to the International Space Station. It became the first commercially built and operated spacecraft to successfully deliver cargo to the ISS in May 2012. After 23 launches, Dragon 1 was retired in 2020.

DRAGON 2

Dragon 2 is now being used, and is able to transport both cargo and crew. Dragon 2 comes in two versions: Crew Dragon and Cargo

Dragon. Cargo Dragon is an updated version of the original Dragon. Both versions are reusable.

SEVEN-SEATER SPACECRAFT

Crew Dragon can fit up to seven astronauts, though NASA missions will only have four astronauts onboard. Not only will Crew Dragon transport astronauts to the ISS, it will also be used to transport paying passengers to and from space.

COST PER ASTRONAUT

After the Space Shuttle Program ended, Russia was the only country sending people into space. At first, Russia sold seats in the Soyuz to NASA for $40 million per astronaut back in 2011. That cost increased over the years to more than $90 million per astronaut by 2020.

In comparison, the cost per astronaut on Crew Dragon costs $55 million. That works out to be $220 million for all four astronauts onboard.[86]

Future Spaceships

THE ORION

The Orion is a partially reusable deep-space spacecraft being built to fly to the Moon—and eventually to Mars. It's named after the Orion constellation. It will consist of a crew module, service module and launch abort system.

Orion Crew Module

The crew module is the habitable part of the spacecraft and can support a crew of six.[87] This is also the piece of the craft that will re-enter Earth's atmosphere, so it has a heat shield that can withstand 5,000°F (2,760°C). It will descend under parachutes and is designed to perform a skip entry splashdown, much like a rock being skipped across a lake. This skip entry will provide a smoother ride, land Orion closer to the U.S. coast and lower the g-forces.[88]

Orion Service Module

The Service Module provided by the European Space Agency is Orion's powerhouse and will provide power, propulsion, thermal control, air and water during the flight to the Moon and back. It has four solar wings that are almost 23 feet (7 meters) long that can rotate and pivot to track the sun. Each solar wing is made up of three solar panels. It will generate and store power while in space.[89]

The Crew Module and Service Module will be attached from launch throughout the entire mission until separating just before re-entry.

Launch Abort System

The Orion also includes a Launch Abort System that can be activated within milliseconds to bring the crew module to safety in case of an emergency. It's also built to steer the crew module for a safe landing. It's built to be 10x safer than the space shuttle.[90]

First Flight Scheduled

Artemis 1 is the first planned unmanned Orion test flight. It is scheduled to take place in February 2022. It will be launched to the Moon, enter lunar orbit and then return to Earth approximately 25 days and 12 hours later. Orion will be launched on the most powerful rocket in the world, the Space Launch System (SLS).[91] Two mannequins will fly onboard Orion and will be used to measure and evaluate the amount of radiation crews could be exposed to during Artemis missions.

SPACEX HUMAN LANDING SYSTEM

In April 2021, NASA announced that SpaceX will build the next lunar lander spacecraft that will work with the Orion spacecraft to land the next humans on the Moon. SpaceX's human landing system, or HLS, will be the spacecraft two crew members use to detach from Orion and land on the Moon. After spending roughly a week on the surface of the Moon, the astronauts will re-board the HLS, return to Orion and begin their trip back to Earth.

The HLS will be part of history not only because we're going back to the Moon, but this time the first woman and first person of color will be aboard.[92]

RUSSIA'S OREL

Orel in Russian means "eagle." This is the next generation of partially reusable spacecraft that will replace the Soyuz spacecraft and will support low Earth orbit and lunar missions. It's comparable to the U.S. Orion spacecraft. Orel will be capable of carrying up to six people (the Soyuz can only carry three) for up to 30-day missions. It can be docked with a space station and could stay docked up to a year, which is double what the Soyuz is capable of. The Orel is scheduled to make its first unmanned test flight to the ISS in 2023, followed by a crewed mission in 2025.[93] Then the plan is to fly to the Moon in 2028.[94]

CHINA'S NEXT SPACECRAFT

China is building a reusable deep-space spacecraft that will carry Chinese astronauts to the Chinese space station and fly to the Moon. It's currently unnamed, but the new spacecraft will replace the current Shenzhou spacecraft and is built to carry a crew of six or a mix of three crew members and 1,102 lbs. (500 kg) of cargo. It has a crew module and service module, like the U.S. Orion and India's Gaganyaan.[95] The first unmanned test flight in May of 2020 was successful. Planned missions to the Moon are expected to launch in the 2030s.[96]

INDIA'S GAGANYAAN

Gaganyaan in Sanskrit means "sky craft." This spacecraft is designed to carry a crew of three and orbit Earth for up to seven days. It will fly automatically and includes a crew module and service module. The Gaganyaan has an emergency escape system and is built for water-landings in the Bay of Bengal. The plan is to launch an uncrewed test flight in December of 2021, and a crewed flight in 2023.[97]

BOEING'S CST-100 STARLINER

Boeing has built a commercial reusable spacecraft that will take professional astronauts to the International Space Station. It can take up to seven passengers or a mix of crew and cargo to the ISS. But for NASA missions, it will carry up to four astronauts. It can be reused up to 10 times, with a break of six months between each mission to get it space-ready. The first crewed test flight and operational crewed flight is planned for 2022.[98]

Boeing built three Starliner spacecraft, but only two will be used and will be rotated for NASA missions. The Starliner was designed to fly on autopilot, but can be flown manually if needed. It was built for land-based landings and comes with airbags.[99]

VIRGIN GALACTIC'S SPACESHIPTWO

Another reusable spaceplane built for space tourism is Virgin Galactic's SpaceShipTwo. Rather than launching vertically from the ground, SpaceShipTwo is carried by WhiteKnightTwo, a four-engine jet aircraft, that flies to around 50,000 feet (15 km). SpaceShipTwo is then released and air-launched into space. When it reaches its maximum altitude of 360,000 feet (110 km), passengers will experience micro-gravity. The spaceship wings then feather as it begins its re-entry glide. It takes roughly 25 minutes to glide back to the spaceport.

SpaceShipTwo can carry six passengers and two pilots. The entire flight will last around two hours.[100] SpaceShipTwo advertises having "more windows than any other spacecraft in history." Tickets were originally sold the price of $250,000, but in August 2021, the price increased to $450,000.[101]

On July 11, 2021, Virgin Galactic launched its first passengers 53 miles above the Earth's surface. Richard Branson was passenger 001 and became the first spacecraft company founder to launch into space[102] on his own spacecraft.

BLUE ORIGIN'S NEW SHEPARD & NEW GLENN

Amazon founder Jeff Bezos owns Blue Origin and has created a suite of reusable spacecrafts to support space tourism, space payloads and moon landings. New Shepard is a fully reusable commercial crew spacecraft to be used for suborbital space tourism flights. It will launch vertically and land vertically, controlled entirely with onboard computers. This means it needs a limited number of ground control assistance and no human pilot. It can carry up to six paying passengers in the Crew Capsule and has a launch escape system. The entire mission will last roughly 11 minutes.[103]

It's named after Alan Shepard, the first American to go to space. The capsule is 10x larger than the Mercury capsule that Shepard flew in. It's large enough to allow passengers to do somersaults and float freely during their zero-g experience. The capsule has several large windows and is advertised as having "the largest windows in space."[104]

On July 20, 2021, on the anniversary of humans landing on the Moon, New Shepard launched into space carrying its first passengers, including Blue Origin founder Jeff Bezos and his younger brother. Also onboard were Wally Funk (age 82) and Oliver Daemen (age 18), who broke the record for being the oldest woman and youngest person in space, respectively.[105]

Blue Origin's New Glenn spacecraft gets its name from John Glenn, the first American to orbit the Earth. New Glenn is a reusable payload spacecraft and is meant to allow civil, commercial and national security customers to release satellites and other orbital payloads.[106]

Blue Origin is also working on Blue Moon spacecraft meant to land on the Moon: one for cargo and one for humans.[107]

DREAM CHASER SPACEPLANE

Two versions of the reusable Dream Chaser spaceplane are being built: a cargo and a passenger version. The cargo version is planned to fly in 2022. It will handle resupply flights to the ISS. It will launch vertically and land horizontally on autopilot. Like the space shuttle, it will glide to Earth after re-entry. But unlike the shuttle, it's small enough that it will be able to land on any airport runway that handles commercial air traffic.

The crewed Dream Chaser will come after the cargo version, and will carry from two to seven people to the International Space Station and back. It will have a built-in launch escape system and will be able to be flown on autopilot.[108]

ESA'S SPACE RIDER

The European Space Agency has plans to develop a reusable spaceplane, called Space Rider. It is meant for uncrewed orbital spaceflight to carry experiments and deploy satellites.[109]

NOTES

1. Barbara Maranzani, "7 Things You May Not Know About John Glenn," History.com, last modified September 24, 2019, https://www.history.com/news/7-things-you-may-notknow-about-john-glenn-and-friendship-7.

2. "Why Was Apollo Called Apollo? The History of Spacecraft Call Signs," Royal Museums Greenwich, accessed May 28, 2021, https://www.rmg.co.uk/stories/topics/why-was-apollo-called-apollo-history-spacecraft-call-signs.

3. "Early Astronaut Selection and Training," NASA, accessed May, 28, 2021, https://science.ksc.nasa.gov/history/early-astronauts.txt.

4. Loyd S. Swenson, James M. Grimwood, and Charles C. Alexander, *This New Ocean: A History of Project Mercury* (Washington, DC: NASA, 1998), 353.

5. Darrell Etherington, "SpaceX Successfully Completes Key Test of Its Crew Dragon Human Spacecraft," TechCrunch, January 19, 2020, https://techcrunch.com/2020/01/19/spacex-successfully-completes-key-test-of-its-crew-dragon-human-spacecraft/.

6. Maranzani, "7 Things."

7. Brian Dunbar, "What Was Project Mercury?" ed. Sandra May, NASA, last modified August 7, 2017, https://www.nasa.gov/audience/forstudents/5-8/features/nasa-knows/what-was-project-mercury-58.html.

8. Brian Dunbar, "What Was the Gemini Program?" ed. Sandra May, NASA, last modified August 7, 2017, https://www.nasa.gov/audience/forstudents/5-8/features/nasa-knows/what-was-gemini-program-58.html.

9. Dunbar, "What Was the Gemini Program?"

10. John Schwartz, "Why Does 'First Man' Say Gemini as 'Geminee'? NASA Explains. Sorta," New York Times, October 17, 2018, https://www.nytimes.com/2018/10/17/movies/first-man-gemini-nasa.html.

11. Brian Dunbar, "What Was the Apollo Program?" ed. Flint Wild, NASA, last modified July 18, 2019, https://www.nasa.gov/audience/forstudents/5-8/features/nasa-knows/what-was-apollo-program-58.html.

12. Emily Kennard, "What's in a Name?" ed. Kathleen Zona, NASA, last modified May 15, 2009, https://www.nasa.gov/centers/glenn/about/history/silverstein_feature.html.

13. "Apollo 7 Flight Journal," NASA, last updated April 2, 2018, https://history.nasa.gov/afj/ap07fj/a7_01_launch_ascent.html.

14. Brian Dunbar, "About Apollo 7, the First Crewed Apollo Space Mission," NASA, last modified October 12, 2018, https://www.nasa.gov/mission_pages/apollo/missions/apollo7.html.

15. Royal Museums Greenwich, "Why Was Apollo Called Apollo?"

16. Royal Museums Greenwich, "Why Was Apollo Called Apollo?"

17. Royal Museums Greenwich, "Why Was Apollo Called Apollo?"

18. Royal Museums Greenwich, "Why Was Apollo Called Apollo?"

19. "Apollo 12 Facts," National Air and Space Museum, accessed September 4, 2021, https://airandspace.si.edu/explore-and-learn/topics/apollo/apollo-program/landing-missions/apollo12-facts.cfm.

20. Eric M. Jones, "That May Have Been a Small One for Neil...," Apollo 12 Lunar Surface Journals, NASA, last modified April 7, 2018, https://www.hq.nasa.gov/alsj/a12/a12.eva1prelim.html.

21. Christopher S. Wren, "Pete Conrad, 69, the Third Man to Walk on the Moon, Dies After a Motorcycle Crash," New York Times, July 10, 1999, https://www.nytimes.com/1999/07/10/us/pete-conrad-69-the-third-man-to-walk-on-the-moon-dies-after-a-motorcycle-crash.html.

22. Craig Glendale, "Guinness World Records, 2010: Farthest Distance from Earth Reached by Humans," Internet Archive (New York: Bantam Books, 2010), https://archive.org/details/guinnessworldrec00vari/page/13/mode/2up.

23. "Odyssey," Merriam-Webster, accessed May 28, 2021, https://www.merriam-webster.com/dictionary/odyssey#note-1.

24. W. David Woods et al., "Apollo 13 Flight Journal – Day 3, Part 2: 'Houston, We've Had a Problem,'" NASA, last modified April 21, 2020, https://history.nasa.gov/afj/ap13fj/08day3-problem.html.

25. Michael Trostel, "3 Things: The Moon Club," United States Golf Association, February 5, 2019, https://www.usga.org/content/usga/home-page/articles/2019/02/three-things-alan-shepard-moon-club.html.

26. Chelsea Gohd, "A Lunar Rock Sample Found by Apollo 14 Astronauts Likely Came from Earth," Astronomy.com, January 29, 2019, https://astronomy.com/news/2019/01/a-moon-rock-collected-by-apollo-14-astronauts-likely-originated-on-earth.

27. Royal Museums Greenwich, "Why Was Apollo Called Apollo?"

28. "The Moon Trees," NASA, accessed May 29, 2021, https://nssdc.gsfc.nasa.gov/planetary/lunar/moon_tree.html.

29. "Apollo 15 Moon Buggy Plaque that Reads 'Man's First Wheels on the Moon, Delivered by Falcon, July 30, 1971,'" Photograph, NASA, accessed July 6, 2021, https://www.hq.nasa.gov/alsj/a15/AS15-88-11862HR.jpg.

30. Donald A. Beattie, Taking Science to the Moon: Lunar Experiments and the Apollo Program (Baltimore, MD: Johns Hopkins Univ. Press, 2003).

31. Eric M. Jones, "Hammer and Feather," Apollo 15 Lunar Surface Journal, NASA, last modified April 19, 2015, https://history.nasa.gov/alsj/a15/a15.clsout3.html.

32. Read the transcript here: https://history.nasa.gov/alsj/a15/a15.clsout3.html

33. See a movie clip (49 sec; 6.2Mb): https://history.nasa.gov/alsj/a15/a15v_1672206.mpg

34. As seen not far from the Lunar Rover in this photo: "Apollo 15 Crew, AS15-88-11902," 1971, photograph, NASA, https://history.nasa.gov/alsj/a15/a15rb11902-3EDH.jpg.

35. Shown in this photo: "Apollo 15 Crew, AS15-88-11901," 1971, photograph, NASA, https://history.nasa.gov/alsj/a15/AS15-88-11901HR.jpg.

36. Shown in this photo: "Charlie Duke, AS16-117-18841," 1971, photograph, NASA, https://www.hq.nasa.gov/office/pao/History/alsj/a16/AS16-117-18841HR.jpg.

37. Royal Museums Greenwich, "Why Was Apollo Called Apollo?"

38. Royal Museums Greenwich, "Why Was Apollo Called Apollo?"

39. "The Blue Marble from Apollo 17," December 7, 1972, photograph, NASA, https://visibleearth.nasa.gov/images/55418/the-blue-marble-from-apollo-17.

40. NASA, "The Blue Marble."

41. "Plaque, Lunar Module, Apollo 17," National Air and Space Museum, accessed May 29, 2021, https://airandspace.si.edu/collection-objects/plaque-lunar-module-apollo-17/nasm_A19751405000.

42. "Encyclopedia Astronautica Apollo 17," Astronautix.com, accessed July 6, 2021, https://web.archive.org/web/20110812193502/http://www.astronautix.com/flights/apollo17.htm.

43. Astronautics.com, "Apollo 17."

44. Johannes Kemppanen, "Apollo Flight Journal – Apollo Lunar Quarantine," NASA, last modified July 22, 2019, https://history.nasa.gov/afj/lrl/apollo-quarantine.html.

45. See the original photo: "Apollo Image Atlas: The Original Blue Marble," Lunar and Planetary Institute, accessed May 29, 2021, https://www.lpi.usra.edu/resources/apollo/frame/?AS17-148-22727.

46. Al Reinert, "The Blue Marble Shot: Our First Complete Photograph of Earth," The Atlantic, April 12, 2011, https://www.theatlantic.com/technology/archive/2011/04/the-blue-marble-shot-our-first-complete-photograph-of-earth/237167/.

47. Kay Grinter, "Space Shuttle Enterprise," John F. Kennedy Space Center, last modified October 3, 2000, https://web.archive.org/web/20150326061638/http://www-pao.ksc.nasa.gov/kscpao/shuttle/resources/orbiters/enterprise.html.

48. "Shuttle Orbiter Columbia (OV-102)," Kennedy Space Center, February 1, 2003, https://science.ksc.nasa.gov/shuttle/resources/orbiters/columbia.html.

49. Brian Dunbar, "Space Shuttle Overview: Columbia (OV-102)," ed. Jeanne Ryba, NASA, last modified April 12, 2013, https://www.nasa.gov/centers/kennedy/shuttleoperations/orbiters/columbia_info.html.

50. Kay Grinter, "Space Shuttle Challenger," John F. Kennedy Space Center, accessed May 29, 2021, https://web.archive.org/web/20090203035705/http://www-pao.ksc.nasa.gov/shuttle/resources/orbiters/Challenger.html.

51. Kay Grinter, "Space Shuttle Discovery," John F. Kennedy Space Center, September 21, 2000, https://web.archive.org/web/20110610033909/http://www-pao.ksc.nasa.gov/shuttle/resources/orbiters/Discovery.html.

52. Brian Dunbar, "Space Shuttle Overview: Atlantis (OV-104)," ed. Jeanne Ryba, NASA, last modified April 12, 2013, https://www.nasa.gov/centers/kennedy/shuttleoperations/orbiters/atlantis-info.html.

53. Elizabeth Howell, "Endeavour: NASA's Youngest Shuttle," Space.com, December 11, 2017, https://www.space.com/18123-space-shuttle-endeavour.html.

54. "125,000 See Endeavour Land: Satellite Rescue Highlights Maiden Trip," Daily Breeze, September 12, 2012, https://www.dailybreeze.com/2012/09/12/125000-see-endeavour-land-satellite-rescue-highlights-maiden-trip/.

55. "20 Years Ago: NASA's Space Shuttle Breaks a World Record for the Longest Time in Space," FAI, December 6, 2016, https://www.fai.org/news/20-years-ago-nasa%E2%80%99s-space-shuttle-breaks-world-record-longest-time-space.

56. NASA, Space Shuttle Era Facts, 2011, pdf, https://www.nasa.gov/pdf/566250main_2011.07.05%20SHUTTLE%20ERA%20FACTS.pdf.

57. Scott Kelly, Endurance: A Year in Space, a Lifetime of Discovery (New York: Alfred A. Knopf, 2017), 210.

58. "STS-3 Columbia Lands at the White Sands Missile Range, NM," NASA, last updated November 1, 2017, https://www.nasa.gov/centers/dryden/multimedia/imagegallery/Shuttle/EC82-18507.html.

59. Hanneke Weitering, "A Soyuz Capsule Just Made a Record-Breaking 3-Hour Flight to the International Space Station," Space.com, October 14, 2020, https://www.space.com/soyuz-makes-fastest-space-station-crew-flight-record.

60. Chris Hadfield, "Spaceships: Navigating to the International Space Station," MasterClass, accessed May 29, 2021, https://www.masterclass.com/classes/chris-hadfield-teaches-space-exploration/chapters/spaceships-navigating-to-the-international-space-station.

61. "Buran Orbiter," Molniya Research Industrial Corporation, accessed May 29, 2021, http://www.buran.ru/htm/molniya.htm.

62. Colin Burgess and Rex Hall, *The First Soviet Cosmonaut Team: Their Lives, Legacy, and Historical Impact* (New York: Praxis, 2009), 100.

63. "Valentina Tereshkova: Whose Will the Woman Who Conquered Space Obeyed," RIA News, June 16, 2006, https://ria.ru/20060616/49619382.html.

64. Sarah Knapton, "Russia Forgot to Send Toothbrush with First Woman in Space," *The Telegraph*, September 17, 2015, https://www.telegraph.co.uk/news/science/space/11871598/Russia-forgot-to-send-toothbrush-with-first-woman-in-space.html.

65. "Call Signs of Astronauts," Cosmonaut Training Center, accessed May 29, 2021, http://www.gctc.ru/main.php?id=156.

66. Royal Museums Greenwich, "Why Was Apollo Called Apollo?"

67. David R. Williams, "Cosmos 110," NASA Space Science Data Coordinated Archive, April 27, 2021, https://nssdc.gsfc.nasa.gov/nmc/spacecraft/display.action?id=1966-015A.

68. Barbara Mattsen, "Luna Series," Goddard Space Flight Center, NASA, February 6, 2014, https://imagine.gsfc.nasa.gov/science/toolbox/missions/luna.html.

69. "Luna Mission," Lunar and Planetary Institute, accessed May 29, 2021, https://www.lpi.usra.edu/lunar/missions/luna/.

70. Asif A. Siddiqi, "Beyond Earth: A Chronicle of Deep Space Exploration, 1958–2016," NASA, accessed May 28, 2021, https://www.nasa.gov/sites/default/files/atoms/files/beyond-earth-tagged.pdf.

71. Charles Fishman, "Guess Who Was Waiting at the Moon for Apollo 11? The Russians," Fast Company, July 18, 2019, https://www.fastcompany.com/90377951/guess-who-was-waiting-at-the-moon-for-apollo-11-the-russians.

72. Marcus Lindroos, ed., "The Soviet Manned Lunar Program," Space Policy Project (Federation of American Scientists), accessed September 4, 2021, https://spp.fas.org/eprint/lindroos_moon1.htm.

73. Becky Little, "The Soviet Response to the Moon Landing? Denial There Was a Moon Race at All," History.com, https://www.history.com/news/space-race-soviet-union-moon-landing-denial.

74. Tim Peake, *Ask an Astronaut: My Guide to Life in Space* (Little, Brown and Company, 2017), 35.

75. Peake, *Ask an Astronaut*, 125.

76. Brian Dunbar, "What Is the Soyuz Spacecraft?" ed. Flint Wild, NASA, last modified June 27, 2018, https://www.nasa.gov/audience/forstudents/k-4/stories/nasa-knows/what-is-the-soyuz-spacecraft-k-4.

77. Samantha Cristoforetti, *Diary of an Apprentice Astronaut*, trans. Jill Foulston (Penguin, 2020), 197, Kindle.

78. Elizabeth Howell, "Russia's Space Traditions! 14 Things Every Cosmonaut Does for Launch," Space.com, June 6, 2018, https://www.space.com/40809-russian-space-launch-traditions.html.

79. Amy Shira Teitel, "Why Cosmonauts Have Never Splashed Down," *Discover Magazine*, November 20, 2019, https://www.discovermagazine.com/the-sciences/why-cosmonauts-have-never-splashed-down.

80. Cristoforetti, *Diary of an Apprentice Astronaut*, 120–121.

81. Peake, *Ask an Astronaut*, 130.

82. Cristoforetti, *Diary of an Apprentice Astronaut*, 130.

83. Cristoforetti, *Diary of an Apprentice Astronaut*, 131.

84. Cristoforetti, *Diary of an Apprentice Astronaut*, 229.

85. Richard Hollingham, "Why Europe's Astronauts Are Learning Chinese," BBC Future, June 27, 2018, https://www.bbc.com/future/article/20180626-why-europes-astronauts-are-learning-chinese.

86. Michael Sheetz, "Why the First SpaceX Astronaut Launch Marks a Crucial Leap for NASA's Ambitions," CNBC, June 3, 2020, https://www.cnbc.com/2020/06/03/first-spacex-astronaut-launch-marks-crucial-leap-for-nasa-ambitions.html.

87. "NASA Orion Flight Test," NASA, accessed May 29, 2021, https://www.nasa.gov/specials/orionfirstflight/.

88. Brian Dunbar, "Orion Spacecraft to Test New Entry Technique on Artemis I Mission," ed. Rachel Kraft, NASA, last modified April 8, 2021, https://www.nasa.gov/feature/orion-spacecraft-to-test-new-entry-technique-on-artemis-i-mission.

89. "Orion's Service Module," NASA, accessed September 3, 2021, pdf, https://www.nasa.gov/sites/default/files/atoms/files/orion_smonline.pdf.

90. Brian Dunbar, "NASA Announces Key Decision for Next Deep Space Transportation System," NASA, August 27, 2017, https://www.nasa.gov/home/hqnews/2011/may/HQ_11-164_MPCV_Decision.html.

91. Brian Dunbar, "Around the Moon with NASA's First Launch of SLS with Orion," ed. Kathryn Hambleton, NASA, February 4, 2020, https://www.nasa.gov/feature/around-the-moon-with-nasa-s-first-launch-of-sls-with-orion.

92. Brian Dunbar, "NASA Picks SpaceX to Land Next Americans on Moon," ed. Katherine Brown, NASA, April 22, 2021, https://www.nasa.gov/press-release/as-artemis-moves-forward-nasa-picks-spacex-to-land-next-americans-on-moon.

93. Vladimir Gubarev, "Roscosmos Decided to Change the Name of the New Spacecraft," Pravda, September 6, 2019, https://www.pravda.ru/news/science/1432996-roskosmos/.

94. "New Oryol Multiple Use Spacecraft to Begin Tests in 2024," TASS Russian News Agency, February 13, 2020, https://tass.com/science/1119601.

95. Andrew Jones, "China Readies Its New Deep-Space Crew Capsule for 1st Test Flight," Space.com, January 23, 2020, https://www.space.com/china-deep-space-crew-capsule-launch-prep.html.

96. Andrew Jones, "This Is China's New Spacecraft to Take Astronauts to the Moon (Photos)," Space.com, October 2, 2019, https://www.space.com/china-new-spacecraft-crewed-moon-missions.html.

97. Surendra Singh, "Gaganyaan Manned Mission Not before 2023: Minister," The Times of India, February 17, 2021, https://timesofindia.indiatimes.com/india/gaganyaan-manned-mission-not-before-2023-minister/articleshow/81013233.cms.

98. Eric Berger, "It Now Seems Likely that Starliner Will Not Launch Crew until Early 2022," Ars Technica, April 13, 2021, https://arstechnica.com/science/2021/04/it-now-seems-likely-that-starliner-will-not-launch-crew-until-early-2022/.

99. "Starliner Design Details," Boeing, accessed May 29, 2021, https://www.boeing.com/space/starliner/#/design-details.

100. Virgin Galactic, "Space and You!" Voyage: Space for Young Explorers, Issue 8 (January 2009), https://bis-space.com/membership/voyage/VoyageIssue8-Virgin%20Galactic.pdf.

101. Joey Roulette, "Virgin Galactic reopens ticket sales starting at $450K per seat," The Verge, August 5, 2021, https://www.theverge.com/2021/8/5/22611847/virgin-galactic-ticket-sales-richard-branson.

102. NASA's definition of space begins at 50 miles above the surface of the Earth.

103. "New Shepard," Blue Origin, accessed May 29, 2021, https://www.blueorigin.com/new-shepard/.

104. "Become an Astronaut," Blue Origin, accessed May 29, 2021, https://www.blueorigin.com/new-shepard/become-an-astronaut/.

105. Elizabeth Howell, "Blue Origin to Launch Its 1st Astronaut Flight with Jeff Bezos and Crew of 3 Today," Space.com July 20, 2021, https://www.space.com/blue-origin-jeff-bezos-first-human-flight-launch-day.

106. "New Glenn," Blue Origin, accessed May 29, 2021, https://www.blueorigin.com/new-glenn/.

107. "Blue Moon," Blue Origin, accessed May 29, 2021, https://www.blueorigin.com/blue-moon/.

108. Brian Dunbar, "Dream Chaser Model Drops in at NASA Dryden," NASA, last modified August 7, 2017, https://www.nasa.gov/centers/dryden/Features/dream_chaser_model_drop.html.

109. "Space Rider: Europe's Reusable Space Transport System," European Space Agency, May 6, 2019, https://www.esa.int/Enabling_Support/Space_Transportation/Space_Rider_Europe_s_reusable_space_transport_system.

TRAINING FOR THE MOON

WHERE ON EARTH DO YOU TRAIN FOR THE MOON?

When NASA was preparing astronauts to land on the Moon, they looked the world over to find places that would best simulate the Moon, its rocks and its craters. NASA sent the astronauts to several places, including Hawaii, Arizona, Texas, Nevada and Iceland. Having multiple and varied geological sites gave the astronauts the experience they needed to feel comfortable identifying valuable samples when they were on the Moon.

EXPLORING CINDER LAKES CRATER FIELD

If you can't find just the right place, create your own! That's what NASA did in order to train the astronauts in a realistic lunar-like landscape. Cinder Lake Crater Field in Arizona is a humanmade crater field that NASA created by blasting the ground with hundreds of pounds of dynamite. They first mapped a portion of the Moon's craters from satellites, including the Sea of Tranquility where Apollo 11 landed.[1] Then they created an exact replica of the craters in a series of carefully ordered blasts.[2] Although the craters aren't as pronounced as they once were due to weathering, they're still visible and can be visited by the public.[3]

During the Apollo years, NASA and its astronauts tested moon rovers, hand tools and the crews' ability to name their location by only looking out of the Lunar Module's windows at the landscape here and then pinpointing it on satellite images.

OLD SCHOOL WAYFINDING IN THE GRAND CANYON

The Apollo astronauts used a classroom setting to learn basic geological principles like how to identify and collect various types of rocks and how to look at the landscape and then identify where you are on a map. Then they got to go on field trips to apply what they learned! They visited the Grand Canyon where they hiked the

South Kaibab Trail to the bottom of the canyon. On their hikes the astronauts had to identify different rocks in the area and then identify their location using satellite images and topographic maps. The next day, they climbed back up using the Bright Angel Trail. The field experience was some of the most valuable training they had to prepare for their moon walks.[4]

COLLECTING ROCKS IN TEXAS
Roughly 90 miles (144 km) southeast of El Paso is Sierra Blanca, where volcanic rocks are scattered about. In February 1969, Apollo 11 astronauts Armstrong and Aldrin had to practice gathering and identifying rock samples. They used tape recorders and cameras to capture their work, which is exactly the same as what they did on the Moon. While it may be easy to collect rocks, it's another thing to accurately name each one, which they were required to do while collecting moon rocks.[5]

EXAMINING ROCKS AT NEVADA'S NUCLEAR TEST SITE
Just east of Death Valley National Park, and 65 miles (104 km) north of Las Vegas, lies the Nevada National Security Site where atomic weapons were tested both underground and aboveground. Some of the underground explosions created massive craters, and a few of these were used as a training ground for the Apollo astronauts. The Sedan and Schooner craters and Buckboard Mesa were considered ideal training sites. The setting and geology of the craters and the nearby volcanic formations like the Timber Mountain caldera helped fine-tune the astronaut's ability to identify and collect samples.[6]

EXPLORING HAWAII'S VOLCANOES
The Hawaiian Islands were created by undersea volcanoes, which is why Hawaii is the perfect place to study different aspects of volcanoes like gas and lava vents, lava lakes, pit craters and more. Between 1965 and 1972, Apollo astronauts spent time on foot and in airplanes to study and observe nine different locations on the Big Island. This included the slopes of Hualālai, Mauna Loa (the world's largest and

most active volcano), Mauna Kea and the variety of craters at the summit of Kīlauea.[7]

TRAINING FOR THE MOON IN ICELAND

Of all the places they used to train the astronauts for the Moon landings, NASA decided Iceland had a strikingly similar landscape to the Moon's lunar landscape. NASA sent 32 astronauts to train just outside Húsavík, a quiet 2,300-person fishing community on Iceland's northern coast. Nine of the 12 men who walked on the Moon, including Neil Armstrong, trained in Húsavík.[8]

Scientists are learning that not only does Iceland share similarities with the Moon, but it has a lot in common with Mars. Iceland's glaciers, volcanoes and hot springs are very similar to how Mars looked billions of years ago. NASA and other space agencies are already testing rovers and equipment—including inflatable equipment—that they hope to use on Mars.[9]

COMPARING ICELAND ROCKS & MOON ROCKS

The basalt rocks and volcanic geology of Iceland's barren highlands look otherworldly and were believed to be the most moon-like. The Apollo astronauts and geologists explored the volcanically active regions of Iceland between July 1965 and July 1967. They practiced picking out the best rocks, identifying them, describing them and using their special moon rock equipment.

PLAYING THE MOON GAME

To help the Apollo astronauts learn to work in pairs and to pick out high quality rock samples from designated areas, Icelandic geologists created a game to improve the Apollo astronauts' geological examination skills. The astronauts would record their reason for selecting certain rocks on handheld tape recorders. The NASA geologists would analyze the rocks and the reasoning for each selection, and award points to each team based on which team collected the widest variety of high quality rock samples.

DID YOU KNOW?

The Exploration Museum in Húsavík was created to teach visitors and the people of Iceland about the role Iceland played in humans walking on the Moon. The museum has a full-sized replica of the Apollo 11 lunar module Eagle, a collection of historical photos and artifacts—like the rocks from the Moon donated to the museum—and Neil Armstrong's crew patch from the Apollo 11 mission, donated by Armstrong's family.

NOTES

1. David A. King, "Lunar Exploration Initiative," Usra.Edu, updated July 2008, https://www.lpi.usra.edu/science/kring/lunar_exploration/CinderLakesCraterField.pdf.

2. quartzcity, "Cinder Lake Crater Field," ed. WhiskeyBristles, Allison, and thenoblewoman. Atlas Obscura, May 2, 2012), https://www.atlasobscura.com/places/cinder-lake-crater-field.

3. Jennifer Nalewicki, "Before Going to the Moon, Apollo 11 Astronauts Trained at These Five Sites," Smithsonian Institution, July 17, 2019, https://www.smithsonianmag.com/travel/going-moon-apollo-11-astronauts-trained-these-five-sites-180972452/.

4. Nalewicki, "Before Going to the Moon."

5. Nalewicki, "Before Going to the Moon."

6. Nalewicki, "Before Going to the Moon."

7. "50 Years After Apollo 11: Remembering Hawaii's Role in the Moon Landings," Pacific International Space Center for Exploration Systems, August 2, 2019, https://pacificspacecenter.com/2019/08/02/50-years-after-apollo-11-remembering-hawai%CA%BBis-role-in-the-moon-landings.

8. Piet van Niekerk and Werner Hoffmann, "Travel – How Iceland Helped Humans Reach the Moon," BBC, July 2, 2019, http://www.bbc.com/travel/gallery/20190701-how-iceland-helped-humans-reach-the-moon.

9. Van Niekerk and Hoffmann, "Iceland."

TO THE MOON & BACK

Getting to the Moon

FAR, FAR AWAY

Most astronauts and cosmonauts only fly an average of 254 miles (407 km) away from Earth to visit the International Space Station. Only 24 men have travelled to the Moon, which on average is a 238,900-mile (384,472-km) trip one-way. The total round trip flight distance for the Apollo 11 crew was 953,054 miles (1,533,791 km), which included 30 orbits around the Moon.[1] To compare, the circumference of the Earth is 24,902 miles (40,070 km).

THE FIRST UNMANNED MOON LANDINGS

Before Americans landed a man to the Moon, the Soviet Union successfully sent a spacecraft there. In fact, they successfully sent three spacecraft to the Moon before Armstrong and Aldrin walked on the Moon.

Luna 2, was an unmanned spacecraft that was intentionally crashed into the Moon's surface on September 13, 1959. It became the first human-made object to reach the Moon.

1966 was a good year for the Soviets. They launched Luna 9 and Luna 10. Both reached the Moon. Luna 9 was the first to successfully reach the Moon's surface with a controlled soft landing. Luna 10 was the first to enter the Moon's orbit, and became the Moon's first satellite.[2]

CREWED MISSIONS TO THE MOON

During NASA's Apollo program from 1961 to 1972, there were nine crewed missions to the Moon and all nine returned safely back to Earth. Three of those missions flew to the Moon, but never landed (Apollo 8, 10 and 13). The other six missions landed men on the Moon.

FLY ME TO THE MOON

Those nine crewed missions flew 24 men to the Moon and back. Three of those men flew to the Moon twice. James Lovell flew onboard Apollo 8 and 13, John Young was on Apollo 10 and 16, and Eugene Cernan flew onboard Apollo 10 and 17.

FIRST TO FLY TO THE MOON

The first men to fly to the Moon were onboard Apollo 8, and flew December 21 to December 27 1968. James (Jim) Lovell, William Anders and Frank Borman made history when they left low Earth orbit, entered the Moon's orbit, circled the Moon 10 times and then flew safely home. They were the first humans to see and photograph the Earthrise.[3]

JOURNEY TO THE MOON

There's a difference between entering lunar orbit and landing on the Moon. On average it took Apollo astronauts just over three days to enter lunar orbit. On day four of their journey, they landed on the surface of the Moon.[4]

THE FIRST TRIP WAS A SHORT STAY ON THE MOON

The entire Apollo 11 mission took 8 days, 3 hours, 18 minutes and 35 seconds from launch to landing. Their historic stay on the Moon lasted just shy of 22 hours.[5]

THE FINAL TRIP TO THE MOON WAS THE LONGEST

Apollo 17 was the final trip to the Moon, and they made sure to make it worth it! The entire mission was 12 days, 13 hours, 51 minutes, 59 seconds. The entire distance traveled was a total of 1,484,933.8 miles (2,389,769 km), which included 75 orbits around the Moon.[6] The final two men on the Moon also spent the longest time on the Moon's surface performing three separate EVA's spending a total of 22 hours and 4 minutes walking on the Moon.[7]

THE MOON LANDING DRESS REHEARSAL

The three men onboard Apollo 10 played a key part in the Apollo missions. Their mission was to do almost everything a crew would do to land on the Moon—except for actually land on the Moon! The astronauts flew to the Moon, undocked the lunar module from the command module and descended to 8.4 nautical miles (15.6 km) above the Moon's surface. They then docked again with the command module and were able to return safely to the Earth.[8]

Two of the three men of Apollo 10, John Young and Eugene Cernan, returned to the Moon and walked on the Moon. The third man on Apollo 10, Thomas Stafford, never returned to the Moon.

APOLLO 13: THE SUCCESSFUL FAILURE

Apollo 13 didn't land on the Moon as intended. When an oxygen tank exploded two days en route, the landing mission was aborted and their mission then became to return back to Earth safely. The crew used the lunar module as a lifeboat, circled the Moon and made their way back to Earth. It was called a successful failure because of what was learned from the mission and because they were able to get the Apollo 13 crew home safely.[9]

I WALKED ON THE MOON

Of the 24 men that flew to the Moon, 12 of those men walked on the surface of the Moon. The only 12 humans to walk on the Moon are, in order:

1. Neil Armstrong (Apollo 11)
2. Buzz Aldrin (Apollo 11)
3. Pete Conrad (Apollo 12)
4. Alan Bean (Apollo 12)
5. Alan Shepard (Apollo 14)
6. Edgar Mitchell (Apollo 14)
7. David Scott (Apollo 15)
8. James Irwin (Apollo 15)

9. John Young (Apollo 16)
10. Charles Duke (Apollo 16)
11. Eugene Cernan (Apollo 17)
12. Harrison Schmitt (Apollo 17)

THE SIX FORGOTTEN APOLLO ASTRONAUTS

The men who didn't walk on the Moon, but stayed on the command module and orbited the Moon while his two crewmates got the glory of walking on the Moon are often forgotten. But they were critical to the success of the Apollo missions. If something had happened to them, then two moonwalkers would have been stranded on the Moon unable to return to Earth. If something happened to the men walking on the Moon and they weren't able to make it back to the Command Module, these pilots would have had to return to Earth alone and would forever be "marked" as leaving men behind, regardless of the circumstances.[10]

Michael Collins wrote this about how he felt when he was alone orbiting the dark side of the Moon: "I am alone now, truly alone, and absolutely isolated from any known life. I am it. If a count were taken, the score would be three billion plus two over on the other side of the Moon, and one plus God knows what on this side." [11]

These men took photographs that were potential future landing sites, performed experiments and are one of the few humans to see the dark side of the Moon.

These are the 6 heroic men who flew to the Moon, but didn't walk on the Moon when their crewmates did:

- Michael Collins – Apollo 11
- Richard Gordon – Apollo 12
- Stuart Roosa – Apollo 14
- Alfred Worden – Apollo 15
- Ken Mattingly – Apollo 16
- Ronald Evans – Apollo 17

DID YOU KNOW?

Half a billion people watched Neil Armstrong take his first steps on the Moon. But one very important person never did: Astronaut Michael Collins. Collins was the third crew member onboard Apollo 11, and he couldn't watch the live broadcast of Neil Armstrong's first steps on the Moon. There wasn't a TV onboard and he was on the far side of the Moon during the time and cut off from all communications.[12]

Apollo 11 – The First Mission to Land on the Moon

APOLLO 11 MISSION PATCH

It's tradition to include the names of the crew on the mission patch, but the crew of Apollo 11 requested that their names be left off. They felt like the first moon landing belonged to everyone. The crew chose to include a bald eagle about to land and holding an olive branch. The patch is very symbolic on purpose. The Apollo 11 Lunar Module was named Eagle and the eagle is an American symbol. The olive branch is a symbol of peace.[13]

THE FIRST WORD SPOKEN ON THE SURFACE OF THE MOON

The first word broadcast back to Earth from the Moon was "Houston." It was the beginning of the radio call from Armstrong and Aldrin to Mission Control in Houston, Texas, to let them know they had safely landed. The full message was: "Houston: Tranquility Base here. The Eagle has landed."[14]

THE FIRST MOONWALK

Apollo 11 astronauts walked on the Moon a little over six hours after they landed. Armstrong and Aldrin's entire EVA lasted two-and-a-half hours. During their time on the surface of the Moon, Armstrong made his famous "One small step for man, one giant leap for mankind" speech, they spoke to President Nixon, took photographs, gathered lunar samples and deployed experiments.[15]

 DID YOU KNOW?
LEFT FOOT FIRST
The first step on the Moon taken by Neil Armstrong was with his left foot.[16]

AFTER ARMSTRONG'S SPEECH

In case the spacewalk was aborted, NASA had instructed Armstrong to collect a small bag of lunar soil, called the contingency sample. He had decided to do this in the sunlight, but first he wanted to get the Hasselblad camera down and mounted on the control unit on his chest. He did that and then snapped the first pictures taken on the surface of the Moon. Armstrong then moved into the sunlight, reached into his thigh pocket and pulled out a collapsible handle with a detachable bag at one end. He managed to scoop up enough moon dust to fill the bag and even got a few small rocks. Then he reported back, "Contingency Sample Mission Accomplished."[17]

WHAT THE SECOND MAN ON THE MOON SAID

Most of us know what the first man said when he stepped on the surface of the Moon, but what about the second man? Buzz Aldrin summed up his initial thoughts of the Earth's moon in four words: "Beautiful view" and "Magnificent desolation."[18]

THE APOLLO 11 MOON DISASTER SPEECH THAT WAS NEVER GIVEN

Thankfully, this speech was never given, but it was written. U.S. President Nixon was prepared to give it if the Apollo 11 astronauts Neil Armstrong and Buzz Aldrin never made it back to Earth. These are the words Nixon's speechwriter, Bill Safire, crafted:

In The Event Of Moon Disaster:
Fate has ordained that the men who went to the Moon to explore in peace will stay on the Moon to rest in peace.

These brave men, Neil Armstrong and Edwin Aldrin, know that there is no hope for their recovery. But they also know that there is hope for mankind in their sacrifice.

These two men are laying down their lives in mankind's most noble goal: the search for truth and understanding.

They will be mourned by their families and friends; they will be mourned by their nation; they will be mourned by the people of the world; they will be mourned by a Mother Earth that dared send two of her sons into the unknown.

In their exploration, they stirred the people of the world to feel as one; in their sacrifice, they bind more tightly the brotherhood of man.

In ancient days, men looked at stars and saw their heroes in the constellations. In modern times, we do much the same, but our heroes are epic men of flesh and blood.

Others will follow and surely find their way home. Man's search will not be denied. But these men were the first, and they will remain the foremost in our hearts.

For every human being who looks up at the Moon in the nights to come will know that there is some corner of another world that is forever mankind.

PRIOR TO THE PRESIDENT'S STATEMENT: The president should telephone each of the widows-to-be.

AFTER THE PRESIDENT'S STATEMENT, at the point when NASA ends communications with the men: A clergyman should adopt the same procedure as a burial at sea, commending their souls to "the deepest of the deep," concluding with the Lord's Prayer.[19]

Behind the Scenes: How Those Men Got to the Moon

400,000 PEOPLE HELPED MEN GET ON THE MOON & BACK

A total of 24 men flew to the Moon and back, but it took a team of over 400,000 people to get them there. These included people at Mission Control like the flight directors, controllers, planners and engineers. They were the people who designed the rockets and who built them. These were managers, supervisors, quality control and safety inspectors. These were programmers, electricians, welders, seamstresses, gluers, painters, doctors, geologists, scientists, trainers, navigators, food scientists and the list goes on.[20]

APOLLO SPACESUITS: MADE BY THE PLAYTEX BRA COMPANY

The spacesuits that were to be used on the Moon needed to be flexible, yet form-fitting. The company that brought America the "Cross Your Heart Bra" in the mid 1950s was the same company that worked with NASA to create the Apollo spacesuits.

About 500 people from NASA and International Latex Corporation (ILC) worked for several years to create the suits that protected the 12 astronauts who walked on the Moon.[21]

CUSTOM-MADE & STITCHED BY HAND

Each spacesuit was tailored to fit each astronaut that would be wearing them. Each astronaut on the Apollo 11 prime crew had three custom-made suits. Two of them were flight ready and one was for training only. The backup crew had a training suit and a backup suit.[22]

SENDING MEN TO THE MOON INVOLVED A LOT OF WORK BY HAND

It's incredible to realize that a lot of what got the astronauts to the Moon and back safely was the result of people building and creating things by hand, not by machine. The heat shield that protected the astronauts as they re-entered the earth's atmosphere was applied to the Apollo spacecraft by hand with a fancy caulking gun. The parachutes that slowed the capsules before they splashed down were sewn by hand and then folded by hand.

There were only three people in the country who were trained and licensed to fold and pack the Apollo parachutes. If something was to happen to them, it could have been devastating to the Apollo program. To minimize the risk, NASA made sure that these three staff members never rode in the same car together in case there was an accident.

PAPER STAR CHARTS

While the Apollo spacecraft did have a computer onboard, the Apollo astronauts would double check the computer's navigation. They took paper star charts with them and a sextant to take star sighting, just like 18th century sailors.

MANUAL TRAJECTORY CALCULATIONS

NASA mathematician Katherine Johnson played a critical role getting Americans into space and to the Moon. Her work at NASA helped calculate the trajectory for the Apollo 11 flight to the Moon. Johnson created a one-star observation system that astronauts used to accurately determine their location.[23] She was key in helping the

Apollo 13 astronauts get home safely with backup procedures and charts.[24]

NEW MATERIAL INVENTED FOR APOLLO SUITS

Most of the Apollo suit's 21 layers were made from material that existed long before men flew to the Moon. But after the fire that killed the crew of Apollo 1 during a test exercise in 1967, NASA requested that the suits be made to withstand temperatures over 1,000°F (537°C). Scientists invented a material made from Teflon-coated glass microfibers, which was called Beta cloth. This was used as the suit's outer layer.[25]

SOLO SPACECRAFTS

These custom-made suits were basically form fitting spacecrafts for one. They kept the Apollo astronauts alive while on the surface of the Moon. These suits had to protect them from fast-moving micrometeoroids, the intense range of temperatures on the Moon that swung from -387°F to 253°F (-232°C to 122°C)[26] and solar radiation. The suits had internal heaters and cooling systems to help them regulate their body temperatures not just from the outside temperatures, but from their own body heat while they worked on the Moon.[27]

PLSS BACKPACKS

The PLSS (Personal Life Support System) Backpacks provided life support to the astronauts during their moonwalks. These life support backpacks could supply the astronauts with air on the surface of the Moon. Four hours' worth for the early Apollo missions and up to 8 hours on Apollo 15 through 17 missions. The packs weighed 84 lbs (38 kg) on Earth, but only 14 lbs (6.4 kg) on the Moon because of its lower gravity. The PLSS pressurized the suit supplied oxygen to breathe, provided cooling, kept suit humidity within safe and comfortable levels and removed carbon dioxide, particulates and odors.[28]

Moon Rocks: The Most Valuable Cargo in History

BRINGING BACK MOON ROCKS

The six Apollo missions that landed on the Moon between 1969 and 1972 brought back a total of 842 lbs (382 kg) of moon rocks, core samples, pebbles, sand and dust from the lunar surface. (That's almost as heavy as a Grand Piano!) There were 2,196 separate samples[29] from six different landing sites on the Moon.[30] NASA broke those samples up into 140,000 pieces to study.

The Soviets also collected samples from three different sites with their unmanned, robotic lunar spacecraft. The spacecraft returned a total of 0.75 lbs (300 grams) of samples, which is still impressive when you realize it was all done by spacecraft on autopilot 238,900 miles (384,400 km) away. Think of it this way: That's 30 Earths away.[31]

MOON ROCK GIFTS

While NASA still has 85% of the Moon rock samples,[32] U.S. President Nixon gifted tiny lunar samples to every U.S. State and territory and 135 countries, along with the recipient's country flag that the Apollo astronauts had taken to the Moon and back.[33]

THE BIGGEST MOON ROCKS

Big Muley is the name of the largest lunar rock collected during the Apollo missions. It was brought back during Apollo 16 and weighs 26 lbs (11.7 kg). It was named after Bill Muehlberger, who was the Apollo 16 geology team leader.[34]

Great Scott is the second largest lunar rock. It weighs 21.2 lbs (9.6 kg) and was brought back during the Apollo 15 mission. This moon rock is named after astronaut David Scott, who was the mission commander during Apollo 15.

Big Bertha was discovered during the Apollo 14 mission. It contains a meteorite from Earth embedded in part of it. This moon rock contains the oldest known Earth rock at around 4 billion years old. It's also the third largest moon rock that was brought back during the Apollo missions. It weighs 19.8 lbs (8.9 kg) and was named after the large German howitzer used in World War I that was also named Big Bertha.

MOON DUST SMELLS!

There is no air on the Moon, but there is still a smell about the Moon. Every astronaut who walked on the Moon noticed it and many said something to Mission Control. Neil Armstrong, the first man to walk on the Moon, described it as "the scent of wet ashes." His crewmate, Buzz Aldrin said it was "the smell in the air after a firecracker has gone off." Astronaut and geologist Harrison Schmitt was part of Apollo 17, the last mission to the Moon. After his second moonwalk he thought it "Smells like someone's been firing a carbine in here."

Unfortunately, the spent gunpowder smell of the Moon dust didn't make it back to Earth for others to experience after it was exposed to air and moisture.[35]

CLINGY MOON DUST

The dust on the Moon would get everywhere and it was sharp! With no water or wind to smooth tiny Moon particles, they were as sharp as glass. It scratched faceplates and camera lenses, destroyed bearings and clogged equipment joints. It would cling to everything like packing popcorn, because lunar dust particles are electrostatically charged.[36]

SLEPT IN THEIR HELMETS AND GLOVES

To keep from breathing in the moon dust and to avoid getting it on their skin, Apollo 11 astronauts Armstrong and Aldrin spent their one night on the Moon sleeping with their helmets and gloves on while in the Lunar Module.[37]

THE ASTRONAUTS STRIPPED NAKED

Apollo astronauts would go out on a lunar EVA in bright, white suits and return a few hours later looking like they just returned from the mines. On Apollo 12, before reentering the command module, both Alan Bean and Pete Conrad stripped naked and stuffed their space suits into a pouch to keep the dust contained.[38]

 DID YOU KNOW?
Apollo astronauts were instructed to only collect lunar samples no larger than their fist.[39] But when they got there and would come across larger moon rocks that looked like they would be of great value to researchers, they would retrieve them anyway!

Items Left on the Moon

MAKING ROOM FOR MOON ROCKS

There was only so much room and only so much weight the Lunar Module could handle. NASA decided that the moon rocks were worth more than whatever they would leave behind. So the strategy was "the more they threw out, the more rocks they could bring back"[40] and that's what they did.

OVER 190 TONS OF STUFF

According to researchers, we left over 190 tons of "cultural material" on the Moon.[41] For perspective, one ton is 2,000 lbs (907 kg)—about the weight of a small car. The third stages of the Saturn V boosters that provided the thrust to get the Apollo astronauts out of Earth's orbit and into Lunar orbit were all decoupled from Command Service Modules and eventually crash-landed into the Moon. They weigh roughly 30,500 lbs (13,834 kg). These are some of the largest and heaviest objects left on the Moon.

Other heavy items left behind include six lunar descent stages weighing 4,931 lbs (2237 kg) and at least four of the six lunar ascent modules, weighing 5,077 lbs or (2,303 kg). NASA's three Moon Buggys brought by Apollo 15, 16 and 17 weighed 460 lbs (210 kg) each. The

Soviet Union/Russia and China both have a few of their own lunar rovers still on the Moon.

For a full list of the bigger and heavier items left on the Moon by the U.S., China, Japan, India and Soviet Union/Russia, see this list here: https://en.wikipedia.org/wiki/List_of_artificial_objects_on_the_Moon

LUNAR MODULES – DESCENT MODULES

The Lunar Modules came in two parts, the Ascent and Descent Modules. The Descent Module was the part that allowed us to use its engines and provide a soft landing on the Moon. It's the part with the spider-like legs. All six Descent Modules are exactly where they were when they safely landed on the Moon.[42]

LUNAR MODULES – ASCENT MODULES

The ascent modules were how the Apollo astronauts were able to leave the surface of the Moon and rendezvous with the Command Module. After that they were no longer needed and were jettisoned, or released. It was deliberate to have them crash-land on the Moon because it would cause an artificial "moonquake" and would be a type of experiment for the seismic equipment left on the Moon to test.

The LMs would either crash land on the Moon, burn up in Earth's atmosphere or in one instance go into orbit around the Sun![43]

However, the location for both Apollo 11 and Apollo 16 lunar ascent modules is currently unknown.

The first two lunar modules were only in test flights and were jettisoned in Earth's orbit and later burned up as they re-entered Earth's atmosphere.

Apollo 10 went to the Moon, but didn't land. When the crew jettisoned their lunar module, named Snoopy, it went into orbit around the Sun. Fifty years after it was released, astronomers are 98% sure that they have located it.[44]

Apollo 13's LM was used as a lifeboat to help return the crew to the Earth after an explosion damaged the Service Module. Once they were close enough to Earth, they jettisoned the LM and it eventually burned up in the Earth's atmosphere.

Apollo 12, 14, 15 and 17 all crash landed on the Moon and their crash location is known to NASA.

Apollo 11 and Apollo 16 are assumed to have crash landed on the Moon, but exactly where, nobody knows. . .[45]

MISSION	LUNAR MODULE	FATE
Apollo 5	Unnamed	Burned up in Earth's atmosphere
Apollo 9	Spider	Burned up in Earth's atmosphere
Apollo 10	Snoopy	In orbit around the Sun
Apollo 11	Eagle	Crash site unknown – assumed to have crash landed on the Moon.
Apollo 12	Intrepid	Crash landed on the Moon
Apollo 13	Aquarius	Used as a 'lifeboat' and later burned up in Earth's atmosphere
Apollo 14	Antares	Crash landed on Moon
Apollo 15	Falcon	Crash landed on Moon
Apollo 16	Orion	Crash site unknown – assumed to have crash landed on the Moon.
Apollo 17	Challenger	Crash landed on the Moon

6 AMERICAN FLAGS & 6 PLAQUES LEFT BEHIND

Each time the Apollo astronauts landed on the Moon, they placed an American flag at their landing site.[46] All six lunar descent modules had plaques with the names of the crew attached to the ladders.

FALLEN ASTRONAUT SCULPTURE & PLAQUE

The crew of Apollo 15 placed a 3.5-inch (8.9-cm) aluminum sculpture in the form of an astronaut in a spacesuit face down in the lunar soil. It was to honor and commemorate the astronauts and cosmonauts who had died in the line of duty. Next to the tiny figurine is a plaque with the names of the 14 men listed alphabetically.

As of 2021, there have been around 38 men and women who have died while involved in space exploration. But in 1971, 14 people were known to have died. These are the names of those men on the plaque:

Fallen Astronaut figurine and plaque.[47] Photo courtesy of NASA.

- Charles Bassett (died Feb. 1966, aircraft accident)
- Pavel Belyaev (died Jan. 1970, disease)
- Roger Chaffee (Jan. 1967, Apollo 1 fire)
- Georgi Dobrovolsky (Jun. 1971 re-entry pressurization failure)
- Theodore Freeman (Oct. 1964, aircraft accident)
- Yuri Gagarin (Mar. 1968, aircraft accident)
- Edward Givens (1967, automobile accident)
- Gus Grissom (Jan. 1967, Apollo 1 fire)
- Vladimir Komarov (Apr. 1967, re-entry parachute failure)
- Viktor Patsayev (Jun. 1971, re-entry pressurization failure)
- Elliot See (Feb. 1966, aircraft accident)
- Vladislav Volkov (Jun. 1971, re-entry pressurization failure)
- Edward White (Jan. 1967, Apollo 1 fire)
- C.C. Williams (Oct. 1967, aircraft accident)

Because it was a space race and secrets were kept, there are two men missing from this plaque. Two cosmonauts from the original selection of 20 cosmonauts had died before Apollo 15: Valentin Bondarenko (Mar. 1961, fire during training) and Grigori Nelyubov (Feb. 1966, train accident/suicide).[48]

2 GOLF BALLS

Alan Shepard, the first American in space, was also the commander of Apollo 14. He made a makeshift golf club and sent two golf balls flying. According to Shepard, his first two swings weren't his best, but the third drove the ball for "miles and miles and miles."[49]

HASSELBLAD CAMERAS

While the film that gave us some of the most famous photos the world has ever seen made it back to Earth, the special cameras used to take those photos were left on the Moon. Weight restrictions forced a choice: Since moon rocks were more valuable than the cameras, there are 12 Hasselblad cameras still on the Moon.[50]

GOLD-PLATED TELESCOPE

The crew of Apollo 16 took a gold-plated extreme ultraviolet telescope to the Moon. They were the first to view planets and stars through a telescope while standing on a body other than the Earth.[51] This telescope not only allowed the observer to see things we could never see from Earth, but also took photos. In total, the astronauts returned with 178 frames of film taken from this special telescope. It was left on the Moon in the Descartes highland region.[52]

106 ITEMS LEFT BY APOLLO 11

There were items left on the Moon by design, some for sentimental and ceremonial reasons and others for necessity. The Lunar Module Descent Stage was designed to stay on the lunar surface. It acted as the launch pad for the Ascent Stage. Other items left by design were experiments, the U.S. flag, the commemorative plaque and items to commemorate lost astronauts/cosmonauts.

The Lunar Legacy Project at New Mexico State University has compiled a detailed list of everything the crew of Apollo 11 left behind and includes a map of where the Toss Zone is.

ITEMS TO HONOR THOSE WHO DIED

To honor the men who died in the fire during the Apollo 1 training exercise on the launch pad, Armstrong and Aldrin left the Mission Patch from Apollo 1 with the names of astronauts Virgil Grissom, Edward White and Roger Chaffee.

Before they died, the Apollo 1 astronauts had three medals made that they were going to fly to the Moon and back with and then present as gifts to their wives. Since the crew of Apollo 1 didn't get a chance to do it themselves, Armstrong and Aldrin took the medals and flew them to the Moon and back for them.[53]

Two medals honoring two cosmonauts who died were also left on the Moon. These medals were for Yuri Gagarin, who was the first man in space and who later died in a training flight, and Vladimir Komarov, died after re-entry when the Soyuz 1 capsule's parachutes failed to open and it crashed into the ground.[54]

The Apollo 11 silicon disc next to a 50-cent coin. Photo courtesy of NASA.

COMMEMORATIVE PLAQUES

The first of six commemorative plaques were left by the Apollo 11 crew attached to the descent stage ladders. This plaque and the last Apollo mission's plaque were unique because rather than just the name of the mission, date and the crew's signatures, they had a special inscription and the U.S. President's signature. It read, "Here men from the planet Earth first set foot upon the Moon July 1969, A.D. We came in peace for all mankind."

The last mission to the Moon, Apollo 17 left a plaque that read, "Here man completed his first explorations of the Moon December 1972, A.D. May the spirit of peace in which we came be reflected in the lives of all mankind."

DISC CARRYING MESSAGES FROM WORLD LEADERS

A silicon disc slightly larger than a fifty-cent coin contains 73 messages of goodwill from world leaders. These messages have been etched into it and are no larger than one-quarter the width of a human hair. A total of 116 countries were contacted and invited to contribute messages, but only 73 responded in time. The disc was placed in a metal case much like a makeup compact container and placed in a white cloth pouch along with other commemorative items and left under the descent stage ladder.[55]

You can read all of the messages and see which countries are represented on the disc as part of the Lunar Legacy Project.[56]

SYMBOL OF PEACE
To symbolize peace, a small gold replica of an olive branch was left where Apollo 11 landed on the Moon.

LEFT-BEHIND EXPERIMENTS
Some items were part of ongoing experiments, like the Passive Seismic Experiment package that measured "moonquakes" and would help scientists understand more about the internal structure of the Moon.[57] The Laser Ranging Retroreflectors helped NASA determine the orientation and orbit of the Moon.[58]

USED AND NO LONGER NEEDED EQUIPMENT
Once an item fulfilled its purpose and was no longer needed, it was thrown out of the lunar module to make room for the precious moon rocks. Parts of their spacesuits used on the Moon were discarded like moon boots (technically the overboot), Portable Life Support System (PLSS) backpacks, armrests, tools they had used to collect lunar samples, tripods, the flag canister, a TV camera, lenses and cables.

2 LARGE & 2 SMALL URINE COLLECTION ASSEMBLIES
Among the items left behind on the Moon were four urine collection devices. It's not clear who wore which, but two were size small and two were size large.[59]

VOMIT AND POO BAGS
Someone was spacesick on the way to the Moon—or maybe both astronauts were sick—otherwise why else would you leave 4 emesis (vomit) bags on the Moon? If they were empty, they'd probably take them for the trip home. Along with the vomit bags, the astronauts also discarded four "defecation collection devices"—aka poo bags![60]

EMPTY FOOD BAGS
Aldrin and Armstrong ate four separate meals during their short stay on the Moon. Their empty food bags were also left behind.

Out of This World Photos

ASTRONAUT-PROOF CAMERAS

NASA engineers purchased a few Hasselblad cameras and made them space-worthy and astronaut-proof. They stripped them down to save weight, and painted them dull black to reduce reflections. To astronaut-proof them they removed the release for the film magazine to prevent it from being accidently bumped in flight.[61]

PHOTOGRAPHY LESSONS

NASA's space photography expert, Dick Underwood, helped NASA and the astronauts understand how to take great photos that would be remembered forever. He said, "Your key to immortality is solely in the quality of your photographs."

He taught the astronauts how to "shoot from the hip" and to

The Earthrise orientation we normally see. Photo courtesy of NASA.

not use the viewfinder. They practiced in the KC-135 aircraft (aka Vomit Comet) to simulate what it would be like to take pictures in microgravity. Underwood had them practice taking photos of their kids at home to help them learn about exposure and camera lens settings. When they went to the Moon, Underwood was confident they'd bring back some great photos.[62]

WHAT THE EARTHRISE PHOTO LOOKED LIKE ORIGINALLY

The famous Earthrise photo taken during the Apollo 8 mission was not taken in the orientation that has been published for the world to see. The familiar image on the left shows the Earth rising above the Moon's surface.

The original picture was taken in this orientation, with the Earth on the left and the surface of the Moon on the right.

Original Earthrise photo orientation taken on Apollo 8.[63] Photo courtesy of NASA.

WHO TOOK THE EARTHRISE PHOTO?

After decades of debate, there is now strong evidence that proves that Apollo 8's astronaut Bill Anders was the one who took the Earthrise photo.[64]

NASA'S MOST REQUESTED PHOTO

One of the most recognizable images around the world and NASA's most requested photo is the one with Buzz Aldrin on the surface of the Moon with Neil Armstrong reflected in Aldrin's visor.[65]

NASA's most requested photo - Buzz Aldrin on the surface of the Moon with Neil Armstrong's reflection in the visor. Photo by Neil Armstrong. Photo courtesy of NASA.

DEVELOPING THE PRECIOUS FILM

Imagine being responsible for developing the film that was used to take photos on another planet! NASA photography expert Dick Underwood and his team came up with a process that would allow them to safely and successfully process the priceless moon footage.

At that time there were machines that could process film in six minutes, but Underwood's process took five hours. Some people complained. But Underwood and his team weren't going to be rushed. They knew they needed to give the priceless film the tender loving processing it deserved. The results were worth every minute.[66]

PROBLEMS IDENTIFYING WHO IS WHO LED TO WEARING STRIPES

No one thought about it until they sat down to identify who was who in the Apollo 11 photos, but there was no way to distinguish who was who when they were wearing identical space suits. The solution was to add a red or yellow stripe on the arm, leg and helmet of the suit of the lead astronaut.

It was too late to add them for the crew of Apollo 12, but from Apollo 13 onward, the Commanders wore stripes and they could identify who was who in the pictures. Astronauts today still wear stripes on their spacesuits for this same reason.[67]

The only image taken of Neil Armstrong on the surface of the Moon. Photo courtesy of NASA.

ONLY ONE IMAGE OF ARMSTRONG

When Apollo 11 astronauts Armstrong, Aldrin and Collins were in quarantine they reviewed the photos they took while on the Moon. Even though there were two professional Hasselblad cameras onboard only one was used, and Armstrong had it for most of the time. Almost all of the high-quality photos are of Aldrin. It wasn't until years later that this one image of an astronaut working at the LM was identified as Neil Armstrong.

TAKING PHOTOS OF EACH OTHER WASN'T ON THEIR SHOT LIST

They traveled all the way to the Moon and there were so many other things to take photos of that it wasn't thought much about. There were no directions to take photos of each other! Following Apollo 11, both astronauts were given cameras and encouraged to take photos of each other.[68]

Photo of Buzz Aldrin's boot print in the lunar soil - Photo by Buzz Aldrin/NASA.[69]

MOON BOOT IMPRESSION SHOT

While Armstrong had the camera most of the time, Aldrin did have the camera for a short time and took the famous photos of the boot impression on the lunar surface. These are photos of Aldrin's bootprints.[70]

 DID YOU KNOW?

12 astronauts walked on the Moon, but only the last two men to walk on the Moon brought their moon boots back to Earth. Astronauts Harrison Schmitt and Gene Cernan of Apollo 17 decided to keep theirs.[71]

Unusual Items Flown to the Moon & Back

MINIATURE LUNAR ROVER LICENSE PLATES

Several miniature lunar rover license plates were created by Boeing—who also made the lunar rovers—and sent with Astronaut Dave Scott aboard Apollo 15. They measured 1.25 by 0.5 inches (32 mm by 20 mm) and were made from lightweight aluminum. They listed the "MOON" as the home state, had 1971 for the year, included registration number "LRV 001" and included the NASA and Boeing logos.

Miniature Lunar Rover license plates made by Boeing. Image by Heritage Auctions, HA.com

The miniature Moon Buggy plates were prepackaged in a pack smaller than a pack of gum. Scott kept them in his left knee spacesuit pocket where they remained during his entire 12-day trip to the Moon and back. He had them on the surface of the Moon during his EVA and while riding the first Lunar Rover.

They have been sold at a number of different auctions throughout the years and have sold for between $2,875 and $34,378. [72]

SPACE SUNGLASSES

Sunglasses are especially important for astronauts. While their helmets had special visors that acted as sunglasses, sunglasses were crucial to protect astronauts' eyes when they were making observations out the capsule windows. The specific sunglasses NASA approved to go to the Moon were called *Original Pilot Sunglasses®* made by American Optical. Each Apollo astronaut carried a pair in one of their many in-flight suit pockets. Each astronaut had their own pair, which means 24 pairs of these sunglasses flew to the Moon and back with 12 of them landing on the Moon. These have been known to sell at auctions starting at $3,520 for an Apollo 9 pair, and on up to $75,000 for a pair that landed on the Moon. [73]

THE $2 AND $20 BILLS FROM APOLLO 15

The crew of Apollo 15 took several $2 and $20 bills with them to the Moon as part of their PPKs or Pilots Preference Kits that contained the astronauts' personal items. While 50 of the $2 bills stayed with Command Module pilot Al Worden as he orbited the Moon, another package of bills went to the surface of the Moon with astronauts Dave Scott and Jim Irwin.

These Moon bills would have been worth a lot more than $2 or $20 when they came back to Earth, but unfortunately that package of bills was left behind by mistake. The Moon-flown bills that did make it back were signed by all three men and were sold along with a signed flight certificate that included the flown bill's serial number.

After 1973, NASA banned this new tradition and no longer allows currency or coins which could be sold at auction to be taken onboard any U.S. space flight. The only space flown bills that are available post-Apollo have been flown onboard a Russian Soyuz or a private commercial flight like SpaceShipOne.[74]

ROBBINS STERLING SILVER MEDALLIONS
Sterling silver medallions made by the Robbins Company in Attleboro, Massachusetts, have been produced for and flown on every mission since Apollo 7. These were paid for by the crews and only NASA astronauts could purchase them at the time. Depending on weight limits for each mission, between 80 and 440 medallions were flown on each Apollo flight.

A total of 2,915 medallions flew onboard Apollo 7 through Apollo 17. Of those, only 1,310 medallions flew onboard the six Apollo missions that landed on the Moon. The backs of the Moon landing mission medallions include the launch, lunar landing and return dates. These are some of the most sought-after lunar collectibles.[75]

RARE APOLLO 11 GOLD ROBBINS MEDALLIONS
Three unique gold medallions were made for the crew of Apollo 11 by the Robbins Company. These rare space-flown gold medallions are one of the few official items that display the original concept that astronaut Michael Collins had for the Apollo 11 mission.

Collins' original idea was to have the eagle carrying an olive branch in its mouth, but NASA thought the eagle's sharp, open talons looked too "warlike," and the olive branch that represented peace was moved to the claws in the official mission patch. The reverse side of the medallion shows the dates of the mission and the crews' surnames.

The Apollo 11 gold medallions have the original design and were taken to the surface of the Moon. All three gold medallions have since been sold to private collectors, with the one belonging to Neil Armstrong being auctioned off for a staggering $2,055,000 in July 2019.[76]

FISHER SPACE PENS
It turns out that these special ball point pens with pressurized ink cartridges can be used in microgravity and in a wide range of

temperatures. NASA ordered 400 of them in December of 1967 and 29 flew to the Moon, with 12 or 13 of them making it to the lunar surface. The rest of the pens were used by the astronauts during their years of training.[77]

DURO MARKER PEN
These lightweight aluminum felt-tip markers made by the Duro Pen Company in New York City were used on all flights after Apollo 8. They were later sold as "The Rocket Astronaut Pen." In total, 27 of these marker pens flew to the Moon and 12 of those made it to the surface of the Moon. Most of them made it back to Earth safely, but one did go for a spacewalk all on its own during Apollo 17's Ron Evans' deep space EVA.[78]

YO-YOS
Apollo 17's Gene Cernan and Harrison Schmitt brought two Yo-Yos to the Moon with them. While they were instructed to leave them, they both brought them back. Cernan's yo-yo was sold at an auction in 2006 for $50,000.[79]

MILLION DOLLAR BACKUP MOON WATCH
While the Apollo astronauts weren't allowed to keep their NASA-issued Omega Speedmaster chronographs, a few astronauts brought their own watches as backups. Apollo 15 astronaut David Scott brought a Bulova stopwatch and Bulova wristwatch as backups. The Bulova stopwatch was used for a few specific tasks and was sold privately in 2011.

However, the Bulova wristwatch proved to be extremely important because the crystal face on the Omega Speedmaster went missing during Scott's second EVA. He used the strap from the Omega Speedmaster on his Bulova wristwatch and wore it over his spacesuit sleeve during his final EVA and continued to use it for the rest of the mission. Scott sold both the Bulova wristwatch and band for $1,625,000 at auction in 2015.[80]

NOTES

1. Sarah Loff, "Apollo 11 Mission Overview," NASA, last modified May 15, 2019, https://www.nasa.gov/mission_pages/apollo/missions/apollo11.html.

2. Asif A. Siddiqi, Beyond Earth: a Chronicle of Deep Space Exploration, 1958–2016 (Washington, DC: NASA History Division, 2018), https://www.nasa.gov/sites/default/files/atoms/files/beyond-earth-tagged.pdf.

3. Brian Dunbar, "Apollo 8," NASA, last modified January 9, 2018, https://www.nasa.gov/mission_pages/apollo/missions/apollo8.html.

4. Dave Roos, "Apollo 11 Timeline: From Liftoff to Splashdown," History.com, last modified December 8, 2020, https://www.history.com/news/apollo-11-moon-landing-timeline.

5. Roos, "Apollo 11 Timeline."

6. Brian Dunbar, "Apollo 17," NASA, last modified January 9, 2018, https://www.nasa.gov/mission_pages/apollo/missions/apollo17.html.

7. Brian Dunbar, "Apollo 17," NASA, last modified January 9, 2019, https://www.nasa.gov/mission_pages/apollo/missions/apollo17.html.

8. Brian Dunbar, "Apollo 10," NASA, June 11, 2019, https://www.nasa.gov/mission_pages/apollo/missions/apollo10.html.

9. Brian Dunbar, "Apollo 13," NASA, January 9, 2018, https://www.nasa.gov/mission_pages/apollo/missions/apollo13.html.

10. David Kerley and Mina Kaji, "50 Years Later: Astronaut Michael Collins on Apollo 11 Mission," ABC News, July 20, 2019, https://abcnews.go.com/Politics/50-years-astronaut-michael-collins-apollo-11-mission/story?id=64228725.

11. "Carrying the Fire Quotes by Michael Collins," Goodreads, accessed May 18, 2021, https://www.goodreads.com/work/quotes/598913.

12. Kenneth Chang, "For Apollo 11 He Wasn't on the Moon. But His Coffee Was Warm," New York Times, July 16, 2019, https://www.nytimes.com/2019/07/16/science/michael-collins-apollo-11.html.

13. Catherine Thimmesh, Team Moon: How 400,000 People Landed Apollo 11 on the Moon (Boston: Houghton Mifflin Harcourt, 2015), 60–61.

14. "Apollo 11 – Technical Air-to-Ground Voice Transcription," NASA, accessed May 18, 2021, https://www.hq.nasa.gov/alsj/a11/a11transcript_tec.html.

15. Loff, "Apollo 11 Mission Overview."

16. Brian Dunbar and Shelly Canright, "Apollo 11 – First Footprint on the Moon," NASA, last modified December 9, 2007, https://www.nasa.gov/audience/forstudents/k-4/home/F_Apollo_11.html.

17. Andrew Chaikin, Man on the Moon: The Voyages of the Apollo Astronauts (Great Britain: Penguin Books, 2019), 209–210.

18. Richard Hollingham, "Apollo in 50 Numbers: The Workers," BBC Future, June 19, 2019, https://www.bbc.com/future/article/20190617-apollo-in-50-numbers-the-workers.

19. Bill Safire, "In Event of Moon Disaster," Archives.gov, July 18, 1969, https://www.archives.gov/files/presidential-libraries/events/centennials/nixon/images/exhibit/rn100-6-1-2.pdf

20. Thimmesh, Team Moon, 4.

21. Thimmesh, Team Moon, 29–30.

22. Rosemary Feitelberg, "Preparing and Dressing Apollo 11's Crew for the New Frontier," WWD, July 19, 2019, https://wwd.com/eye/people/preparing-and-dressing-apollo-11s-crew-for-the-new-frontier-1203223358/.

23. Meghan Bartels and Paulina Cachero, "Katherine Johnson, the NASA Mathematician Who Helped Put a Man on the Moon, Died at 101. Here's a Look Back at the Hidden Figure's Remarkable Life," Business Insider, February 24, 2020, https://www.businessinsider.com/katherine-johnson-hidden-figures-nasa-human-computers-2016-8.

24. Tim Ott, "Katherine Johnson," Biography.com, last modified January 11, 2021, https://www.biography.com/scientist/katherine-g-johnson.

25. Andrew Chaikin, "Neil Armstrong's Spacesuit Was Made by a Bra Manufacturer," Smithsonian Institution, November 1, 2013, https://www.smithsonianmag.com/history/neil-armstrongs-spacesuit-was-made-by-a-bra-manufacturer-3652414/.

26. Brian Dunbar, "The Moon," NASA, June 25, 2020, https://www.nasa.gov/moon.

27. Tim Sharp, "What Is the Temperature on the Moon?" Space.com, October 27, 2017, https://www.space.com/18175-moon-temperature.html.

28. Eric M Jones, "Apollo Lunar Surface Journal: PLSS (Portable Life Support System)," NASA, August 14, 2016, https://history.nasa.gov/alsj/plss.html.

29. Meghan Bartels, "The Moon on Earth: Where Are NASA's Apollo Lunar Rocks Now?" Space.com, July 14, 2019, https://www.space.com/where-are-nasa-apollo-moon-rocks.html.

30. "Lunar Rocks and Soils from Apollo Missions," NASA, accessed May 19, 2021, https://curator.jsc.nasa.gov/lunar/index.cfm.

31. "How Far Away Is the Moon?" NASA, September 30, 2019, https://spaceplace.nasa.gov/moon-distance/en/.

32. Bartels, "The Moon on Earth."

33. "Where Today Are the Apollo 17 Goodwill Moon Rocks?" collectSPACE.com, accessed May 19, 2021, http://www.collectspace.com/resources/moonrocks_goodwill.html.

34. Eric M Jones, "Station 1 at Plum Crater Transcript and Commentary," NASA, last modified April 7, 2018, https://www.hq.nasa.gov/alsj/a16/a16.sta1.html.

35. Charles Fishman, "What You Didn't Know About the Apollo 11 Mission," Smithsonian Institution, last updated June 19, 2019, https://www.smithsonianmag.com/science-nature/what-you-didnt-know-about-apollo-11-mission-fifty-years-ago-180972165/.

36. Mary Roach, Packing for Mars: The Curious Science of Life in the Void (New York: WW Norton, 2011), 180.

37. Fishman, "What You Didn't Know."

38. "Apollo 12 Mission Log," Earth to the Moon, accessed May 19, 2021, http://www.earthtothemoon.com/apollo_12.html.

39. Jones, "Station 1 at Plum Crater."

40. Anna Heiney, "Apollo's Lunar Leftovers," ed. Jeanne Ryba, NASA, last modified November 22, 2007, https://www.nasa.gov/missions/solarsystem/f_leftovers.html.

41. Minerva Baumann, "NMSU Professor Emerita and Co-Authors Launch Space Preservation Book," NMSU News Center, April 3, 2017, https://web.archive.org/web/20170404025122/https://newscenter.nmsu.edu/Articles/view/12431/nmsu-professor-emerita-and-co-authors-launch-space-preservation-book.

42. "Location of Apollo Lunar Modules," National Air & Space Museum, Smithsonian Institute, accessed May 19, 2021, https://airandspace.si.edu/explore-and-learn/topics/apollo/apollo-program/spacecraft/location/lm.cfm.

43. Tom Housden et al., "Apollo 11: Four Things You May Not Know about the First Moon Landing," BBC News, July 13, 2019, https://www.bbc.com/news/science-environment-48907836.

44. "Snoopy Lunar Module from 1969 Apollo 10 Mission May Have Been Found," Sky News, June 9, 2019, https://news.sky.com/story/snoopy-lunar-module-from-1969-apollo-10-mission-may-have-been-found-11738299.

45. David R. Williams, "Apollo – Current Locations," NASA, accessed May 19, 2021, https://nssdc.gsfc.nasa.gov/planetary/lunar/apolloloc.html.

46. Clara Moskowitz, "Apollo Moon Landing Flags Still Standing, Photos Reveal," Space.com, July 27, 2012, https://www.space.com/16798-american-flags-moon-apollo-photos.html.

47. "Memorial to Fallen Astronauts on the Moon," NASA, February 6, 2019, https://solarsystem.nasa.gov/resources/2279/memorial-to-fallen-astronauts-on-the-moon/.

48. Rose Eveleth, "There Is a Sculpture on the Moon Commemorating Fallen Astronauts," Smithsonian Institution, January 7, 2013, https://www.smithsonianmag.com/smart-news/there-is-a-sculpture-on-the-moon-commemorating-fallen-astronauts-358909/.

49. Heiney, "Apollo's Lunar Leftovers."

50. Michael Zhang, "There Are 12 Hasselblad Cameras on the Surface of the Moon," PetaPixel, June 15, 2011, https://petapixel.com/2011/06/15/there-are-12-hasselblad-cameras-on-the-surface-of-the-moon/.

51. Heiney, "Apollo's Lunar Leftovers."

52. Elizabeth Landau, "Remembering the First Moon-Based Telescope," ed. Tricia Talbert, NASA, last modified August 9, 2019, https://www.nasa.gov/feature/remembering-the-first-moon-based-telescope.

53. "U.S. Taking Russian Medals to the Moon," Chicago Tribune, July 18, 1969, https://chicagotribune.newspapers.com/clip/29967920/chicago-tribune/.

54. "U.S. Taking Russian Medals."

55. Robert Z. Pearlman, "The Untold Story: How One Small Disc Carried One Giant Message for Mankind," Space.com, November 16, 2007, https://www.space.com/4655-untold-story-small-disc-carried-giant-message-mankind.html.

56. "Archeological Inventory at Tranquility Base," Lunar Legacy Project, New Mexico State Univ., accessed May 19, 2021, https://spacegrant.nmsu.edu/lunarlegacies/artifactlist.html.

57. Walter S Kiefer, "Apollo 11 Passive Seismic Experiment," USRA Lunar and Planetary Institute, accessed May 19, 2021, https://www.lpi.usra.edu/lunar/missions/apollo/apollo_11/experiments/pse/.

58. Brian Dunbar, "The Apollo 15 Lunar Laser Ranging RetroReflector," NASA, April 2, 2019, https://www.nasa.gov/mission_pages/LRO/multimedia/lroimages/lroc-20100413-apollo15-LRRR.html.

59. "Archeological Inventory at Tranquility Base."

60. "Archeological Inventory at Tranquility Base."

61. Scott Neuman, "The Camera That Went to the Moon and Changed How We See It," NPR, July 13, 2019, https://www.npr.org/2019/07/13/735314929/the-camera-that-went-to-the-moon-and-changed-how-we-see-it?t=1609842966325.

62. Thimmesh, Team Moon, 50.

63. Bill Anders, "Earthrise – Apollo 8," Flickr, December 24, 1968, https://www.flickr.com/photos/nasacommons/9460163430/.

64. Andrew Chaikin, "Who Took the Legendary Earthrise Photo From Apollo 8?" Smithsonian Institution, January/February 1, 2018, https://www.smithsonianmag.com/science-nature/who-took-legendary-earthrise-photo-apollo-8-180967505/.

65. Thimmesh, Team Moon, 49.

66. Richard W. Underwood, "Oral History," interview by Summer Chick Bergen, NASA Johnson Space Center Oral History Project, October 17, 2000, https://historycollection.jsc.nasa.gov/JSCHistoryPortal/history/oral_histories/UnderwoodRW/UnderwoodRW_10-17-00.htm

67. Eric M Jones, "Commander's Stripes," Apollo Lunar Surface Journal, NASA, February 20, 2006, https://www.hq.nasa.gov/alsj/alsj-CDRStripes.html.

68. Neuman, "Camera That Went to the Moon."

69. Buzz Aldrin, "Aldrin's Boot Print on Moon," National Air and Space Museum, Smithsonian Institution, July 20, 1969, https://airandspace.si.edu/multimedia-gallery/5302hjpg.

70. Buzz Aldrin, "Bootprint on the Lunar Surface," NASA, accessed May 19, 2021, https://www.nasa.gov/mission_pages/apollo/40th/images/apollo_image_11a.html.

71. Eric M Jones, "Post-EVA-3 Activities, Corrected Transcript and Commentary," Apollo 17 Lunar Surface Journal, NASA, last modified September 30, 2014, https://www.hq.nasa.gov/alsj/a17/a17.eva3post.html.

72. Chris Spain, "Apollo 15 Miniature LRV License Plates," Space Flown Artifacts, accessed May 19, 2021, http://www.spaceflownartifacts.com/flown_license_plates.html.

73. Chris Spain, "Flown Sunglasses," Space Flown Artifacts, accessed May 19, 2021, http://www.spaceflownartifacts.com/flown_sunglasses.html.

74. "Apollo 15," The Jefferson Space Museum, accessed May 19, 2021, http://www.jeffersonspacemuseum.com/apollo-15

75. Chris Spain, "Apollo-Era Robbins Medallions," Space Flown Artifacts, accessed May 19, 2021, http://www.spaceflownartifacts.com/flown_robbins_medallions.html.

76. "Neil Armstrong's One and Only Apollo 11 Lunar Module Flown MS67 NGC: Lot #50067," Heritage Auctions, accessed September 7, 2021, https://historical.ha.com/itm/explorers/space-exploration/neil-armstrong-s-one-and-only-apollo-11-lunar-module-flown-ms67-ngc-14k-gold-robbins-medal-directly-from-the-armstrong-fam/a/6209-50067.s.

77. Chris Spain, "Flown Writing Instruments from the Mercury, Gemini & Apollo Missions," Space Flown Artifacts, accessed May 19, 2021, http://www.spaceflownartifacts.com/flown_writing_instruments.html.

78. Spain, "Flown Writing Instruments."

79. Jones, "Post-EVA-3 Activities Transcript."

80. Chris Spain, "Timepieces," Space Flown Artifacts, accessed May 19, 2021, http://www.spaceflownartifacts.com/flown_timepieces.html.

THE VERY FIRST ASTRONAUTS

These are the first five humans to risk everything in the name of space exploration and strap into tiny, cramped capsules on top of repurposed military missiles[1] with the goal of making it to space and back safely.

YURI GAGARIN – THE FIRST HUMAN IN SPACE

On April 12, 1961, 27-year-old cosmonaut Yuri Gagarin made history when he launched into space onboard the Vostok spacecraft and orbited Earth. He wasn't just the first human to fly into space he was also the first to orbit Earth. It was his first and only flight.

The Soviet Union selected a group of 20 men to train and prepare to fly into space. These men needed to be pilots, weigh less than 159 lbs (72 kg) and be 5 ft 7 in (1.70 meters) or smaller, since the Vostok spacecraft was anything but roomy. Of the original 20, six were selected and were called the "Sochi Six." These were the men who would make up the first cosmonauts of the Vostok program. Gagarin was one of those six.

Gagarin was 5 ft 2 inches (1.57 meters) tall, was favorited among his peer group and was described by the Air Force doctor to be: "Modest; embarrasses when his humor gets a little too racy; high degree of intellectual development evident in Yuri; fantastic memory; distinguishes himself from his colleagues by his sharp and far-ranging sense of attention to his surroundings; a well-developed imagination; quick reactions; persevering, prepares himself painstakingly for his activities and training exercises, handles celestial mechanics and mathematical formulae with ease as well as excels in higher mathematics; does not feel constrained when he has to defend his point of view if he considers himself right; appears that he understands life better than a lot of his friends."[2]

Gagarin's entire flight from launch to landing lasted 118 minutes. His Vostok spacecraft left the pad at 6:07 GMT. The Vostok spacecraft separated from the rocket and began its orbit at 6:17 GMT. At 7:25 GMT, the spacecraft began its deorbit burn. At 7:35 GMT, the spacecraft began its re-entry. At 7:55 GMT, Gagarin ejected from his capsule and deployed his parachute, and 10 minutes later at 8:05 GMT he was safely on the ground. His total time in spaceflight was 1 hour and 18 minutes, or 78 minutes.[3]

He returned home a national hero, and everything Gagarin did on his first launch (including peeing on the back-right tire of the bus on his way to the launchpad[4]) have become traditions that cosmonauts and astronauts honor to this day every time they fly from Baikonur Cosmodrome.

Unfortunately, Gagarin died at the age of 34 on a routine training flight on March 27, 1968.

His historic accomplishment is celebrated every year on April 12 in Russia on what's known as Cosmonautics Day. In 2011, the United Nations declared April 12 as International Day of Human Space Flight. On April 12, people all over the world celebrate with festivities called "Yuri's Night."

ALAN SHEPARD – FIRST AMERICAN & SECOND HUMAN IN SPACE

Just weeks after Gagarin made his historic flight, NASA launched Alan Shepard onboard the Freedom 7 on May 5, 1961. Shepard became the first American in space and the second man in space. His entire 15-minute suborbital flight was broadcast live for the world to see.[5]

VIRGIL GRISSOM – THIRD HUMAN IN SPACE

Virgil "Gus" Grissom was the third human and second American in space. He flew a little over two months after Shepard on July 21, 1961 onboard the Liberty Bell 7. His suborbital flight lasted 15 minutes and 37 seconds.[6]

GHERMAN TITOV – FOURTH HUMAN IN SPACE

In August of 1961, Gherman Titov became the second person to orbit Earth and the fourth human in space onboard the Vostok 2. He was just under 26 years old when he flew to space and still holds the record for being the youngest person to fly in space. He set many space firsts during his 17.5 orbits of earth over his 1 day, 1 hour and 18-minute flight. He was the first to take photos of and film the Earth, first to sleep in space and the first to suffer from space sickness.

JOHN GLENN – FIRST AMERICAN TO ORBIT THE EARTH

On Feb. 20, 1962, astronaut John Glenn became the fifth human in space and the first American to orbit the Earth onboard Friendship 7. He orbited the Earth three times over the course of almost 5 hours.

Almost Astronauts

LOVELACE'S WOMEN IN SPACE PROGRAM

The NASA doctor who was responsible for developing mental and physical tests to ensure astronaut candidates could handle spaceflight was Dr. Randy Lovelace. He put the original Mercury 7 men through all of their testing and he believed that women were just as capable as men.

He also believed that women could save the space program a lot of money. They were generally smaller and lighter than men, which meant they use less oxygen and take up less room in the small Mercury spacecrafts. He figured that female astronauts could save NASA nearly $1,000 per pound.

To prove that women were just as capable, Loveless selected a group of 13 women who were highly qualified pilots to put through the same tests as the Mercury 7 men. The program was done privately by Lovelace and privately funded. These women, who later became known as the Mercury 13, all passed the physical tests. Additional testing involved space simulation and that required the use of military facilities. The U.S. government wouldn't allow Lovelace to use military equipment to test women. So before they could finish the testing, the Lovelace program was canceled and NASA kept the test pilot restrictions in place that kept women from being able to become early astronauts.

FIRST LADY ASTRONAUT TRAINEES

These are the names of the 13 women who were highly skilled and talented pilots and who were part of Lovelace's tests: Geraldyn "Jerrie" Cobb, Myrtle Cagle, Janet Dietrich, Marion Dietrich, Wally Funk, Sarah Gorelick, Jane "Janey" Briggs Hart, Jean Hixson, Rhea Woltman, Gene Nora Stumbough, Irene Leverton, Jerri Sloan and Bernice Steadman.

Only one woman completed all of the tests and was in the top 2% of all candidates for both males and females. Geraldyn "Jerrie" Cobb even outperformed some of the Mercury 7 astronauts.

GERALDYN "JERRIE" COBB

Cobb was a skilled pilot who loved to fly. She earned her solo pilot's license at age 16, and by age 20 she had both her private and commercial pilot's licenses. While still in her 20's she set three separate world records in aviation. In 1959, she set the world record for nonstop long-distance flight and the world light-plane speed record. In 1960, she set the world altitude record for lightweight aircraft of 37,010 feet (11, 290 meters). She was the first woman to fly at the prestigious Paris Air Show and was awarded the Amelia Earhart Gold Medal of Achievement. She logged more than 7,000 hours in the air—more than John Glenn's 5,000 hours and Scott Carpenter's 2,900 hours.[7]

Even with her impressive world records, awards and extensive flight experience, Cobb was denied any chance of becoming an astronaut. She passed all the training exercises and ranked in the top 2% of all astronaut candidates of both genders. Cobb left NASA for the Amazon jungle and spent four decades delivering food, medicine and other aid to the indigenous people as a solo pilot.[8]

NASA'S MILITARY TEST PILOT REQUIREMENT

Because NASA decided to make it a requirement to be a military test pilot, this eliminated women from applying to be an Apollo astronaut. At the time, women were banned from flying in the military after the end of the Women Air Force Service Pilots (WASP) in 1944.[9]

CHANGE IN REQUIREMENTS

It wasn't until the Space Shuttle program came along that NASA decided they needed two types of astronauts: pilots and mission specialists. Pilots could be military or civilian. Test pilot qualifications were no longer required, but were preferred. Mission specialists would do everything else except pilot or command the space shuttles. Mission specialists could be civilian engineers, physicians and scientists. This change is what made it possible for women and people of color to apply and be accepted as astronaut candidates.[10]

MAKING IT TO SPACE, FINALLY

Pilot and flight instructor Wally Funk has 19,600 hours of flight experience and has taught over 3,000 people to fly. Funk is one of the original 13 women of the Mercury 13, and she finally got her chance to fly into space alongside Amazon's Jeff Bezos and his brother onboard Blue Origin's New Shepard in 2021.[11] On July 20, 2021, the anniversary of humans landing on the Moon, she broke two world records. At the age of 82, she beat Peggy Whitson's record of being the oldest woman in space and John Glenn's record of being the oldest human in space.

NOTES

1. "What Is the Difference between Rocket and Missile?" Smithsonian National Air and Space Museum, Smithsonian Institute, April 20, 2017, https://howthingsfly.si.edu/ask-an-explainer/what-difference-between-rocket-and-missile.

2. "Yuri Gagarin," European Space Agency, accessed May 24, 2021, https://www.esa.int/About_Us/ESA_history/50_years_of_humans_in_space/Yuri_Gagarin.

3. Robert Frost, reply to "When Yuri Gagarin went into space, the journey took 108 minutes. Now I am confused whether he spent 108 minutes in space or did the whole journey take 108 minutes?" Quora, April 5, 2020, https://qr.ae/pNQIeg.

4. Ella Morton, "Why Russian Astronauts Pee on a Bus Tire Before Launching Into Space, and Other Pre-Flight Rituals," Atlas Obscura, July 29, 2015, https://www.atlasobscura.com/articles/why-russian-astronauts-pee-on-a-bus-tire-before-launching-into-space-and-other-preflight-rituals.

5. "Mission Monday: Five Fast Facts about the First American in Space," Space Center Houston, May 4, 2020, https://spacecenter.org/mission-monday-five-fast-facts-about-the-first-american-in-space/.

6. Mary C. White, "Detailed Biographies of Apollo I Crew – Gus Grissom," NASA, last modified August 4, 2006, https://history.nasa.gov/Apollo204/zorn/grissom.htm.

7. "Geraldyn 'Jerrie' M. Cobb Collection Gawey," National Air and Space Museum, accessed September 6, 2021, https://airandspace.si.edu/collection-archive/geraldyn-jerrie-m-cobb-collection-gawey/sova-nasm-2020-0018.

8. Swapna Krishna, "The Mercury 13: The Women Who Could Have Been NASA's First Female Astronauts," Space.com, July 24, 2020, https://www.space.com/mercury-13.html.

9. Alessandra Potenza, "We Fact-Checked Mercury 13, Netflix's Doc about NASA's First Women Astronaut Trainees," *The Verge*, May 29, 2018, https://www.theverge.com/2018/5/29/17393698/netflix-documentary-mercury-13-women-space-astronauts-margaret-weitekamp-interview.

10. National Research Council, *Preparing for the High Frontier: The Role and Training of NASA Astronauts in the Post- Space Shuttle Era* (Washington, DC: National Academies Press, 2011): 23, https://www.nap.edu/read/13227/chapter/3#23.

11. "Wally Funk Will Fly to Space on New Shepard's First Human Flight," YouTube (blueoriginchannel, 2021, https://www.youtube.com/watch?v=lDyakSKpBmU.

ASTRONAUT/ COSMONAUT FIRSTS

Women Astronaut Firsts

The first man in space was Yuri Gagarin, who left earth on April 12, 1961. It wasn't until June 16, 1963 that the first woman was launched into space. At that time, 11 men had already been to space before the first woman. The Russians were the first to send a man and a woman to space. Americans didn't send a woman into space until June 18, 1983, when Sally Ride flew onboard the Space Shuttle *Challenger*. NASA sent a total of 117 men to space before Americans sent Sally Ride.

VALENTINA TERESHKOVA – FIRST WOMAN IN SPACE

Cosmonaut Valentina Tereshkova was 25 years old when she became the first and the youngest woman in space on June 16, 1963 aboard the Vostok 6. She was the 12th person in space and is the only woman to have been to space on a solo mission. As the Vostok 6 took off, she called out "Hey sky, take off your hat. I'm on my way!" Valentina's call sign was "Seagull" and she orbited the earth 48 times in just under three days.

Before she was recruited as a cosmonaut in 1962, Tereshkova worked in a textile factory and was an amateur skydiver. Even though she wasn't a pilot, she had 126 parachute jumps under her belt, which is why she was accepted into the program. She was one of 400 applicants and only four women selected to train for 18 months with the hope to eventually fly on Vostok 6. She was the only one of the four women to fly into space.

Because of the space race, being selected and training to fly into space was kept a secret. Her mother didn't even know she was flying into space—she only found out after Tereshkova returned home!

Before she left for her space flight training, Tereshkova had told her mother that she was going for special parachute training.[1]

During her space flight, Tereshkova realized the spacecraft wasn't descending, but only ascending due to a programming error. This could have been disastrous, but she let the engineers know and they were able to correct the problem and get her safely home.

Tereshkova's flight wasn't easy. The capsule was cramped and she suffered from space sickness throughout her flight. The ground crew forgot to pack her toothbrush, but she was resourceful and used her fingers and water to clean her teeth during her almost three-day flight.[2] When she ejected from the spacecraft after re-entry, she hit her head and ended up with a massive bruise on her face.[3]

Tereshkova returned home a national hero and was given the title "Hero of the Soviet Union," and a crater on the Moon and an asteroid are named after her.[4] She never flew to space again, but became a test pilot and earned her doctorate in aeronautical engineering. She also trained future cosmonauts at the Yuri Gagarin Cosmonaut Training Center.

> "A bird cannot fly with one wing only. Human space flight cannot develop any further without the active participation of women." – **Valentina Tereshkova**[5]

FIRST WOMAN TO SPACEWALK & SECOND WOMAN IN SPACE

Nineteen years after the first woman was launched into space, Russians sent a second woman into space: Cosmonaut Svetlana Savitskaya. Savitskaya was also the first woman to fly to space twice and the first woman to walk in space. Not many people realize that she's also the *only* Russian woman to walk in space as of April 2021. Her spacewalk lasted 3 hours and 35 minutes.[6]

Savitskaya was a skilled skydiver who had made 450 parachute jumps by the age of 17. She was also an acrobatic and test pilot with multiple records in supersonic and turboprop aircrafts. She set three world records in space diving from the stratosphere from 45,275 and

46,751 feet (13,800 and 14,250 meters) and another 15 world records in jumps from jet planes.[7] While the Soviet Union/Russia was the first country to send a woman into space and the first to give a woman a spacewalking mission, there have only been a total of four women out of 121 Russians sent to space.[8] Russia recently announced that their only female cosmonaut, Anna Kikina, has been assigned to a mission to the International Space Station in 2022.[9] Her mission will include a spacewalk.

FIRST AMERICAN WOMAN IN SPACE

After 22 years and 117 men had been sent to space, NASA finally sent an American woman into space. Dr. Sally Ride, a physicist, was chosen as part of a group of 35 who were selected from a group of 8,000 applicants in 1978. Five other women were also chosen, but it was Ride who was selected to be the first to fly to space five years later.

Ride was 32 when she flew into space for the first time. As of February 2021, she is the youngest American astronaut to fly into space. She made two trips, and spent a total of 14 days and 7 hours in space.

When she wasn't in space, Ride was training and working behind the scenes developing the robotic arm that would deploy and retrieve satellites. She later successfully used this same robotic arm on her first shuttle mission. Ride also worked as CAPCOM in ground control communicating with astronauts during the second and third shuttle missions. She served on both committees that investigated the Space Shuttle *Challenger* and *Columbia* disasters. She is the only person to have served on both committees.[10]

After she retired from NASA, Ride co-founded Sally Ride Science with her long-term partner of 27 years, Tam O'Shaughnessy. Sally Ride Science is a company that creates entertaining science programs and publications for young students—especially girls.

Ride was a talented and nationally ranked tennis player. She pursued a professional tennis career at one point in college. She earned her Ph.D. in physics and co-wrote several space books with O'Shaughnessy that encourage children to study science and space. Ride is the first known LGBT astronaut.

*"If we want scientists and engineers in the future, we
should be cultivating the girls as much as the boys."*
— **Sally Ride**

FIRST JEWISH WOMAN & SECOND AMERICAN WOMAN IN SPACE

Dr. Judith Resnik was one of the first women to die on a space mission during the *Challenger* disaster, along with American teacher Christa McAuliffe. Resnik was an electrical engineer, software engineer, biomedical engineer and pilot. She was one of the original six women NASA chose as astronaut candidates in 1978. Resnik was trained along with Sally Ride to be the first American woman in space. Ride was ultimately chosen.

Resnik became the fourth woman in space and the second American woman in space on her first space shuttle mission in August 1984 on the maiden voyage of space shuttle *Discovery*. She had helped develop the software and the operating procedures for the shuttle's robotic arm and she was an expert in using it.

Resnik earned a perfect score on her SATs as a teenager, was a classical pianist and gourmet cook.[11] While completing her Ph.D. in engineering she qualified as a professional aircraft pilot. Her father has said, "She had the brain of a scientist and the heart of a poet."[12]

FIRST AMERICAN WOMAN TO SPACEWALK & EXPLORE THE DEEPEST OCEAN

Kathryn D. Sullivan was one of the other five women selected as astronaut candidates alongside Sally Ride in 1978. She made history when she became the first American woman to step outside the airlock on space shuttle *Challenger* on October 11, 1984. Her 3.5-hour spacewalk with fellow astronaut David Leestma involved showing that a satellite could be refueled in space. Sullivan has been to space three times, was involved in many space-advancing experiments and was involved in deploying the Hubble Space Telescope.[13]

Besides being the first American woman to spacewalk, Sullivan set three Guinness World Records[14] including:

- First woman to reach the Challenger Deep
- Greatest vertical extent travelled by an individual within Earth's exosphere
- First person to visit space and the deepest point on Earth

These three records were set in June 2020 when she dove to the Challenger Deep in the Mariana Trench which is the deepest part of the Earth's oceans.[15]

THE FIRST BRITISH ASTRONAUT
Helen Sharman set a lot of firsts on her trip to space including being the first Briton, first European and first British Cosmonaut in space.

Her journey to space was very different than that of professional astronauts and cosmonauts. In 1989 while driving home from work, Sharman heard a radio ad calling for applicants to be the first British Space Explorer who would fly to space as part of Project Juno. This was a joint project between private British companies and the Soviet Union. Out of nearly 13,000 applicants, 26-year-old Helen was chosen to be the first British astronaut.

Before Sharman could start her 18 months of intense training in Star City, Russia, she was required to learn Russian. She would be flying into space with Russian cosmonauts in a Russian spacecraft and spending time on the Russian space station Mir, so knowing how to read and communicate in Russian was critical.

Sharman and two Russian cosmonauts launched onboard the Russian Soyuz TM-12 spacecraft on May 18, 1991. At the age of 27, she became the first woman to visit the space station Mir. While she was there, she conducted medical, chemical and agricultural experiments, spoke to British schoolchildren over amateur radio and observed and photographed parts of the UK.[16]

SHARMAN'S SPACE PANSY SEEDS
Helen brought 125,000 pansy seeds with her to space and then brought them back to Earth for British school children to use as part of a UK-wide experiment to explore the effects of space travel

on the seeds. The space seeds were distributed in packets of 25 to schools all over Britain. The schools also received a batch of control seeds so they could compare the growth of each.[17]

> *"Looking at the Earth from space made me realise the interconnectedness between everything and everyone on that tiny, fragile and precious blue dot. I realised that, despite our differences, we are all living in the same boat (or spaceship). I felt a sense of responsibility towards our home, to the life and lives it supports. This feeling has stayed with me ever since. Among astronauts, like me, we call this feeling the Overview Effect."* – **Helen Sharman**[18]

FIRST LONG SPACE DURATION FLIGHT BY A WOMAN

Russian Cosmonaut Elena Vladimirova Kondakova spent five months on a long-duration mission onboard space station Mir from 1994 to 1995. She was the first woman to complete a long-duration mission in space.

FIRST AFRICAN-AMERICAN WOMAN ASTRONAUT

Mae Jemison has a background in chemical engineering and is a medical doctor. She also became the first black woman astronaut. Jemison flew onboard Space Shuttle *Endeavour* in 1992 and conducted several important experiments dealing with weightlessness and motion sickness. She was one of 15 candidates chosen from about 2,000 applicants in 1987.

 DID YOU KNOW?
Mae Jemison is the first real-life astronaut to appear on *Star Trek* when she made an appearance as Lieutenant Palmer in the episode, "Second Chances" on *Star Trek: The Next Generation*.[19]

FIRST CANADIAN WOMAN & NEUROLOGIST IN SPACE

Canada sent their first woman into space when Roberta Bondar flew in 1992 on the space shuttle. She not only became the first Canadian woman in space, but she was also the first neurologist to visit space. Bondar was chosen as one of the first six original astronauts in Canada in 1983. Her work studying the effects of the human body in low-gravity situations has helped NASA prepare astronauts for long duration missions.

Bondar holds a Bachelor of Science degree in zoology and agriculture, is an accomplished and published nature and landscape photographer, a certified sky diver and private pilot and enjoys hot air ballooning.[20]

FIRST HISPANIC WOMAN IN SPACE & DIRECTOR OF JOHNSON SPACE CENTER

Dr. Ellen Ochoa holds a Ph.D. in Electrical Engineering, is an accomplished flutist and has been to space four times. She was the first Hispanic woman in space, and also the first Hispanic director and second woman director of the Johnson Space Center in Houston, Texas.

Throughout her career, Ochoa has played a major role in the advancement of computerized optical systems that automatically analyze details about objects that they "see." These include patenting an optical system that detects defects in repeating patterns. She was also the co-inventor on three patents for an optical inspection system, an optical recognition method and a method for removing visual noise from images.[21]

FIRST JAPANESE WOMAN IN SPACE

Dr. Chiaki Mukai holds two doctorate degrees—one in medicine and one in physiology—and is a high-ranking cardiovascular surgeon. She also became the first Japanese woman to fly into space and the first Japanese astronaut to fly to space twice. Her first flight was in 1994 where she spent 15 days onboard the space shuttle *Columbia*. Her second flight was in 1998 onboard space shuttle *Discovery*.

During her space career, Mukai has been able to help with many medical experiments related to the cardiovascular system, space biology, radiation biology,[22] and other space medical experiments looking at the impact of space travel on the body and the aging process.[23]

FIRST FRENCH WOMAN IN SPACE

French cosmonaut, Dr. Claudie Haigneré became the first and currently only French woman to fly into space. She was selected as one of seven—and the only woman—to become an astronaut candidate in the French Space Agency in 1985 out of 700 applicants.[24]

She was selected to be part of the Franco-Russian Cassiopée mission as a Research Cosmonaut and trained in Russia. Haigneré launched onboard a Soyuz and spent 16 days on space station Mir. During this trip, she also became the first woman to qualify as a Soyuz Return Commander which meant that she could command a three-person Soyuz crew in the event of an emergency. She also qualified as an engineer and emergency pilot in the space shuttle.

On her second mission, Haigneré became the first European woman to visit the International Space Station in 2001.[25]

FIRST INDIAN-ORIGIN WOMAN IN SPACE

Dr. Kalpana Chawla was born in Karnal, India, and was the first woman ever to enroll in the aeronautical engineering course at the Punjab Engineering College in India. She continued her studies in the U.S. and earned a Ph.D. in aerospace engineering. Chawla's research was published in several technical journals.

After she became a naturalized U.S. citizen, Chawla applied to become a NASA astronaut. She was selected as an astronaut candidate in 1995 and her first flight was in 1997 onboard the Space Shuttle *Columbia*, where she became the first woman of Indian origin to go to space. She flew a second time in 2001 onboard the space shuttle *Columbia*. She was part of the crew who died when *Columbia* disintegrated during re-entry.[26]

FIRST CHINESE-BORN WOMAN IN SPACE

Major Liu Yang is an experienced pilot with over 1,680 hours of flying. She trained for two years to become an astronaut and became the first Chinese woman in space on June 16, 2012 onboard the Shenzhou 9. She goes by the name "little Flying Knight" on China's Tencent QQ messaging service. She loves to cook, read and is an eloquent speaker.[27]

> "International cooperation is very necessary. Chinese have a saying, 'When all the people collect the wood, you will make a great fire.'" – **Liu Yang**

FIRST WOMAN TO COMPLETE A LONG DURATION MISSION

Yelena Kondakova was the third Soviet/Russian female cosmonaut in space and the first woman to make a long-duration spaceflight. She was also the first woman to enter the cosmonaut program with male classmates. Until then, all cosmonaut candidate programs were gender-specific, not mixed.

Kondakova has been to space twice. Her first trip was onboard a Soyuz in 1994 and she spent 169 days (5 months) onboard the Mir space station. Her second trip to space was onboard the Space Shuttle *Atlantis* in May of 1997. She is the first female to fly onboard both an American and Russian spacecraft. She has a degree in mechanical engineering where she specialized in aircraft production.[28]

FIRST ITALIAN WOMAN IN SPACE

Samantha Cristoforetti is a former Italian Air Force fighter pilot and has a degree in mechanical engineering. Cristoforetti has flown six types of military aircrafts and has over 500 hours of flight experience. She is the second astronaut in space to have been to Space Camp.[29] She can speak Italian, English, German, French, Russian and is currently learning Mandarin. During her first mission to the International Space Station, she became the first female ESA astronaut on a long-duration mission[30] and set the record for the longest uninterrupted spaceflight by a European astronaut for a total of 199 days, 16 hours and 42 minutes.[31]

Cristoforetti has been assigned to return to the International Space Station a second time sometime in 2022 and is currently in training.[32]

FIRST WOMAN TO PILOT AND COMMAND THE SPACE SHUTTLE

Colonel Eileen Collins wasn't just the first woman to pilot a space shuttle (Feb. 1995), she was also the first female space shuttle commander (July 1999). She flew to space four times and was the commander for the space shuttle twice and the pilot twice.

On her fourth and last shuttle flight, STS-114, the first shuttle flight after the *Columbia* disaster, Collins performed the first Rendezvous Pitch Maneuver roughly 600 feet below the space station before docking. This was a 360-degree pitch maneuver that allowed ISS astronauts to check for any damage that could cause issues during re-entry.[33]

Collins was an Air Force test pilot and instructor pilot. She holds two Masters degrees; one in operations research from Stanford University and the second in space systems management from Webster University.[34]

FIRST RUSSIAN WOMAN TO VISIT THE ISS

Cosmonaut Yelena Serova is the first female cosmonaut to visit the ISS and the fourth female cosmonaut. She spent 167 days onboard the ISS, and returned home on March 11, 2015. Serova has since retired as a cosmonaut.

FIRST WOMAN COMMANDER OF THE ISS

Commander Peggy Whitson is NASA's most experienced astronaut to date, with 10 spacewalks under her belt. She has spent more time in space than any other American and any other woman in the world. Not bad for a woman who had applied to become an astronaut five times![35] Whitson kept trying and after ten years of applying, she was selected as an astronaut in 1996. She's gone on to set so many records in space exploration it's hard to keep track of.[36]

Whitson was the first female commander of the ISS. They liked her so much the first time they gave her the job a second time, making her the first female astronaut to command the ISS twice. Whitson has been on three long-duration space missions. During her second mission she was made ISS Commander in April 2008. On her third and last mission in 2017, she commanded the ISS for a second time. Even though Whitson has only been to space three times, she holds the record for the most time spent in space by an American as well as a woman of any nationality.[37]

Whitson served as Chief of the Astronaut Corps from October 2009 to July 2012 and became the first female, nonmilitary person to do so.[38]

Whitson grew up on a farm in Iowa and sold chickens to earn the money to get her private pilot's license. She earned several degrees, including her Ph.D. in Biochemistry. She was named NASA's first Science Officer onboard the ISS.[39]

FIRST WOMAN TO LAUNCH ABOARD A SPACEX DRAGON

Dr. Shannon Walker became the first woman to launch into space onboard the SpaceX Dragon named *Resilience*, as part of the first crew. She has a bachelor's degree in physics and a Master of Science degree and a Ph.D. in space physics. Before she was an astronaut, Walker worked at the Johnson Space Center and was a robotics flight controller and flight commander in the Mission Control Center for several shuttle missions.[40]

FIRST AMERICAN WOMAN TO LIVE ABOARD MIR

Dr. Shannon W. Lucid was one of the first six women to be selected as NASA astronauts in 1978. Lucid was a mother when she was selected to be an astronaut, which made her the first mother to be chosen as an astronaut. She spent 188 days onboard the space station Mir. This made her the first American woman to live onboard Mir and the first American woman to complete a long-duration mission. She launched on March 22, 1996 onboard Space Shuttle *Atlantis* and returned home on September 26, 1996.[41] She's been to space five times and at one time held the record for the longest duration stay in space by an American.

For her mission to Mir, she was awarded the Congressional Space Medal of Honor and became the first woman to receive that award. The Russian President Boris Yeltsin also awarded her with the highest Russian award that can be given to a noncitizen, the Order of Friendship Medal.[42]

Lucid is a commercial, instrument and multi-engine rated pilot.[43] She earned her Ph.D. in biochemistry and after her five space missions, she served as the Chief Scientist of NASA from 2002 to 2003. She was also the lead CAPCOM for several shuttle missions.

FIRST ALL-WOMAN SPACEWALK

The first all-female spacewalk happened just a few years ago on Oct. 18, 2019. NASA astronauts Christina Koch and Jessica Meir were able to repair a faulty battery component outside the International Space Station. Koch was the lead astronaut on the EVA and this was the first spacewalk for Meir.[44]

LUCKY LIFE SUPPORT BACKPACK

Astronaut Christina Koch wore the same portable life support system, or PLSS (pronounced pliss), backpack that Kathy Sullivan, the first American to spacewalk, did 35 years ago in 1984.[45]

FIRST MOTHER IN SPACE

Dr. Anna Fisher was one of the first six women selected in NASA Astronaut Group 8 in 1978, along with Sally Ride and Kathryn Sullivan. After being selected as an astronaut, she married another astronaut from NASA Astronaut Group 8, Dr. William Fisher. She flew her first and only shuttle mission in November of 1984 onboard *Discovery*. She had her first daughter in July of 1983. Afterwards, she took some time off from NASA to raise her family.

Fisher is a chemist and emergency physician. She was the physician in the rescue helicopters for the space shuttle flight tests, STS 1 through 4. Fisher helped develop and test the shuttle's robotic arm, contributed to the spacesuit designs to fit women (extra-small EMUs) and has held several management positions in the Astronaut Office. She's been involved in three NASA programs: the Shuttle, International Space Station and the upcoming Orion Project.[46]

FIRST LGBT ASTRONAUTS

After Sally Ride died from cancer in 2012, her obituary shared that her partner of 27 years was her childhood friend, Tam O'Shaughnessy.[47] This news made the public aware that Sally Ride was the first known LGBT astronaut and the first one to fly into space twice.

Since then, there have been two more openly LGBT astronauts to fly to space—astronaut and Navy helicopter pilot Wendy Lawrence and astronaut Anne McClain. Lawrence flew to space four times on the space shuttle and is a retired Navy Captain. Lawrence was the first female Navy helicopter pilot to fly into space.[48]

Lieutenant Colonel Anne McClain is a U.S. Army helicopter pilot and has been to the International Space Station as Flight Engineer for Expedition 58/59 and is one of the Artemis astronauts who could fly to the Moon in coming years.[49]

MORE FIRSTS FOR WOMEN ASTRONAUTS

NASA has said they want to send the first woman to the Moon in 2024. Of the 18 astronauts announced for the Artemis Moon missions, nine of them are women.[50]

In April 2021, the United Arab Emirates announced that 27-year-old Nora Al Matrooshi will become the first female Arab astronaut. She has a degree in mechanical engineering and was chosen from over 4,000 candidates.[51]

Anna Kikina, a 29-year-old engineer and economist, is Russia's only current female cosmonaut. She has been assigned to fly into space in 2022, which will make her the fifth Russian woman in space. She is expected to perform a spacewalk during her mission, which will make her the second female cosmonaut to spacewalk and the first to spacewalk after the Soviet Union's breakup in 1991. To date, Valentina Tereshkova, Svetlana Savitskaya, Yelena Kondakova and Yelena Serova have been the only women cosmonauts.[52]

First Men on the Moon

NEIL ARMSTRONG

Neil Armstrong learned to fly an airplane and flew solo before he had a driver's license. He was a naval aviator who flew 78 missions over Korea. He earned his Master of Science degree in aerospace engineering. After Apollo 11, Armstrong taught aerospace engineering at the University of Cincinnati for eight years. He was an excellent pilot and he proved that when he manually piloted the lunar module out of a rocky lunar crater and maneuvered it to safety with only 20 seconds worth of fuel left!

Armstrong has said that while he had thought about what his first words would be after stepping on the surface of the Moon, he didn't decide what to say until after they landed. While he and Aldrin were still in the LM, Armstrong decided to say his now-famous line, "That's one small step for man. One giant leap for mankind." [53]

He was famously shy around people and intensely private, but in 1985 Armstrong was part of a group of the "greatest explorers" to journey to the North Pole. This group included many famous explorers, including Edmund Hillary, who was one of the first climbers to summit Mt. Everest. Armstrong said he had wanted to see what the North Pole looked like from the ground, since he'd only ever seen it from the Moon. [54]

BUZZ ALDRIN

Dr. Buzz Aldrin was an engineer, fighter pilot and the first astronaut with a doctoral degree. He earned a Doctorate of Science in astronautics and was one of the first two people to land on the Moon in July 1969. After they landed—but before they ventured out onto the surface of the Moon—Aldrin used a kit given to him by his pastor and took communion and read from the New Testament. The ceremony was kept secret due to a lawsuit over the reading of Genesis on Apollo 8. He is the first and only person to hold a religious ceremony on the Moon.

Aldrin's first words after he set foot on the Moon were "Beautiful view." Armstrong responded with "Isn't that something? Magnificent sight out there." Buzz responded with "Magnificent desolation."

Aldrin flew to space twice during his career as an astronaut. Once onboard Gemini 12 where he performed three EVAs. His second and last flight to space was to the Moon on Apollo 11, where he had his fourth EVA on the surface of the Moon as the second man on the Moon.[55]

MICHAEL COLLINS

Michael Collins was the third astronaut onboard Apollo 11, and was the astronaut who flew the command module and had to watch his crewmates land on the surface of the Moon with the lunar lander. Collins was the one who had to worry about returning to Earth alone in case something happened to Neil and Buzz. While his crewmates were taking their historic steps on the Moon, Collins orbited the Moon, experienced the dark side of the Moon and lost radio signal with Mission Control for 48 minutes out of every orbit and was completely alone. He wasn't lonely during his 22 hours alone in orbit, though.[56] He had hot coffee, could listen to the music he wanted to and had the command module all to himself with an epic view.[57]

 DID YOU KNOW?

The first words broadcast from the surface of the Moon were not spoken by Neil Armstrong and they were not, "Houston, Tranquility Base here. The Eagle has landed." The very first words broadcast from the surface of the Moon were from Buzz Aldrin from inside the lunar module, who said, "Contact Light," to alert Mission Control that one of the probes attached to one of the lunar module legs had touched the surface of the Moon.

The second words, most likely when all lunar module legs had touched down on the surface of the Moon, were "Ok, Engine stop," also spoken by Aldrin. Watch the full Apollo 11 descent on YouTube to see and hear for yourself.

Other Historic Astronaut Firsts

FIRST MAN TO SPACEWALK

Cosmonaut Alexei Leonov was the first human to leave the safety of his spacecraft and perform a spacewalk, or EVA, on March 18, 1965. During his spacewalk, Leonov filmed Earth and reported that "The Earth is absolutely round." He was tethered to the Voskhod spacecraft with a cord 16 feet (5 meters) long. His spacewalk lasted 12 minutes and 9 seconds. His suit swelled up due to the pressure, but he was able to release the pressure in his suit and safely reenter the airlock.

Part of the reason the Soviet Union chose Leonov was that he was a talented artist. He brought colored pencils and paper onboard so he could draw his view of the sunrise over the earth. So not only was he the first to spacewalk, he was the first to draw in space.[58]

FIRST AMERICAN TO SPACEWALK

On June 3, 1965—just over 2 months after the Soviets—an American performed a spacewalk during the Gemini 4 flight. Astronaut Ed White became the first American to leave the safety of the spacecraft and perform an EVA. This was a key step in becoming ready to send astronauts to the Moon and to walk on the Moon.

FIRST MEN TO FLY TO THE MOON

The first men to leave low Earth orbit and to orbit the Moon were NASA astronauts Jim Lovell, Bill Anders and Frank Borman on Apollo 8 over Christmas 1968.

FIRST MAN TO ORBIT THE MOON ALONE

John Young, as part of Apollo 10, was the first human to fly solo around the Moon. At the same time, his crewmates flew the Lunar Module down to within 50,000 feet (15.6 km) of the lunar surface as a test run for Apollo 11. Young was alone on the command module for 8 hours until the Lunar Module docked again with the Command Module. Young noticed the number of craters on the far side of the Moon. He said, "Most of the backside of the Moon is just highland impacts." [59]

FIRST AFRICAN AMERICAN IN SPACE

Guy (Guion) Bluford was the first African American in space when he flew on the first of his four space shuttle flights in August 30, 1983. His first spaceflight, STS-8, was the first to launch and land both at night. He was the part of the 1978 NASA Astronaut Group 8 that included the first group of six women, three African Americans and one Asian American. He is an aerospace engineer, fighter pilot and an officer in the U.S. Air Force.

FIRST AFRICAN AMERICAN TO WALK IN SPACE

Dr. Bernard A. Harris, Jr. is a medical doctor and a flight surgeon. He joined NASA as a clinical scientist and flight surgeon where he researched how humans adapt to weightlessness and how best to counteract the negative effects. He was the first African American man to go into space as a part of NASA's research teams. During his second space shuttle flight in February 1995, he performed his first and only spacewalk and became the first African American to do so.[60]

FIRST AFRICAN AMERICAN TO PILOT AND COMMAND THE SHUTTLE

Fred Gregory was one of the three original African American astronauts chosen in 1978 as part of the NASA astronaut Group 8. He flew on three shuttle missions and was the first African American shuttle pilot and shuttle commander. He was a military engineer, Air Force pilot, test pilot and worked at NASA as Deputy Administrator for a time. He has over 7,000 flight hours in more than 50 types of aircrafts, including 550 combat rescue missions in Vietnam.[61]

FIRST AFRICAN AMERICAN TO LIVE ON THE ISS

Victor Glover has three Master of Science degrees and is a test pilot in the U.S Navy. One of his first commanding officers gave him the nickname "Ike," which now has become his call-sign. It stands for "I Know Everything." [62] He has over 3,000 flight hours in more than 40 aircrafts and was the pilot on the first operational flight of the SpaceX

Crew Dragon.[63] He was the first African American to be assigned on an ISS expedition, where he spent over 6 months onboard the ISS.[64]

FIRST CANADIAN ASTRONAUT

Dr. Marc Garneau earned his Ph.D. in Electrical Engineering and was a Captain in the Canadian Forces Maritime Command. In 1983, he was one of the original six Canadian astronauts chosen from over 4,000 applicants—and the only military officer. Garneau flew on three shuttle missions, worked as CAPCOM for several shuttle missions and logged over 677 hours in space.[65]

FIRST CANADIAN ASTRONAUT TO WALK IN SPACE

Chris Hadfield is a pilot and has a degree in mechanical engineering. He has been to space three times, twice on the space shuttle and once onboard a Russian Soyuz. He became the first Canadian to perform a spacewalk on his second flight in April 2001. During his first spacewalk, the anti-fog solution used to polish his spacesuit visor caused major eye irritation and blinded him temporarily while in space. He was able to vent the oxygen into space which fixed the issue and he carried on with his spacewalk. On his third trip to space, Hadfield became the first Canadian to command the International Space Station. He has spent 166 days in space and has performed two spacewalks. He also acted as chief CAPCOM for 25 shuttle missions! [66]

FIRST AFGHAN COSMONAUT

Abdul Ahad Mohmand spent nine days onboard the Mir space station in 1988 and became not only the first Afghan cosmonaut in space, but the first to speak the Pashto language during a conference call to the President of Afghanistan and his mother during his mission. While in space, Mohmand took photos of his country, performed a variety of experiments and brewed Afghan tea for the crew.[67] Before becoming a cosmonaut he had been trained as a pilot in the Soviet Union. After returning home, he moved to Germany and became an accountant.

FIRST BRAZILIAN IN SPACE

Marcos Pontes was the first Brazilian and the first native Portuguese speaker in space on March 30, 2006. He is an experienced Brazilian Air Force pilot with over 2,000 hours in 25 different aircrafts. He was selected by the Brazilian Space Agency to train with NASA to fly on the space shuttle. He completed the training with NASA, but due to budget concerns and the *Columbia* Space Shuttle disaster was unable to fly on the shuttle. He switched to train with Russia on the Soyuz. Pontes spent 8 days in space onboard the International Space Station after launching into space on a Russian Soyuz.[68]

FIRST HISPANIC ASTRONAUT

Dr. Franklin Chang-Díaz earned a Ph.D. in plasma physics and is tied for the record for the most spaceflights. He has seven space shuttle missions under his belt. He was born in Costa Rica and moved to the U.S. as a teenager. He became the first naturalized U.S. citizen to become an astronaut, in 1981.[69] Since then, he's led the research for plasma propulsion to take humans to space. Chang-Díaz currently works as a space entrepreneur as CEO and chair of Ad Astra Rocket Company.[70]

FIRST CUBAN, LATIN AMERICAN & BLACK COSMONAUT

Arnaldo Tamayo Méndez has accomplished many firsts in his life. After he became the first Cuban citizen, Latin American/Caribbean and first person of African descent in space as a cosmonaut in September 1980, he became the first to be awarded the Hero of the Republic of Cuba medal. He was also awarded the Hero of the Soviet Union. Before becoming a cosmonaut, Méndez was a fighter pilot and Chief of Staff at the Santa Clara Aviation Brigade.[71]

FIRST PERUVIAN ASTRONAUT

Carlos I. Noriega was five when he moved from Peru to the United States with his family. He has degrees in computer science and space systems operations. He has flown Sea Knight helicopters and has worked as an instructor pilot and aviation safety officer with the U.S.

Marine Corps. He's been to space twice—the first time in 1997—and has performed three separate spacewalks to help install the first set of U.S. solar arrays on the International Space Station.[72]

FIRST ICELANDIC-BORN ASTRONAUT

Bjarni Tryggvason was born in Reykjavik, Iceland, but grew up in Vancouver, British Columbia in Canada. He has experience as a pilot, pilot instructor, aerobatic pilot and meteorologist and also has a degree in engineering physics. He was one of the original six Canadian astronauts chosen in 1983. He developed systems used on the International Space Station, NASA and the Mir space station.[73] He flew onboard the Space Shuttle *Discovery* in 1997 for a 12-day mission focused on studying changes in Earth's atmosphere.[74]

FIRST SPANISH ASTRONAUT

Pedro Duque has flown on both the space shuttle and the Russian Soyuz and has spent a total of 18 days, 18 hours and 46 minutes in space. He is the first Spaniard to have flown in space when he flew onboard Space Shuttle *Discovery* in October 29, 1998.[75] In October 2003, he spent ten days onboard the ISS. He worked as crew support on Europe's Columbus ISS laboratory and the Cupola module on the International Space Station.

FIRST SYRIAN ASTRONAUT

Muhammed Faris was a pilot in the Syrian Air Force and flew as a Research Cosmonaut in the Interkosmos program onboard a Soyuz to the Mir space station in 1987. He was the first Syrian and second Arab in space, and the first person to have carried dirt from Earth into space. The soil was from Damascus, the capitol of Syria, where he is from.[76]

FIRST ITALIAN IN SPACE

Dr. Franco Malerba has two doctorate degrees: one in electronical engineering and the second in physics. He was selected as an astronaut by the Italian Space Agency and NASA. He flew onboard

Space Shuttle *Atlantis* in 1989 on mission STS-46, where he was in charge of the Italian Tethered Satellite instruments. In his spare time, he enjoys mountaineering, skiing and tennis. He's fluent in Italian, English and French.[77]

FIRST CHINESE TAIKONAUT

Yang Liwei is the first Chinese taikonaut sent to space by the Chinese space program. He was selected as an astronaut candidate in 1998 and flew into space onboard the Shenzhou 5 spacecraft on October 15, 2003. He spent 21 hours, 22 minutes and 44 seconds in space and enjoyed packets of shredded pork with garlic, Kung Pao chicken and eight treasure rice. He washed these meals down with Chinese herbal tea.

After his flight, Yang became a national hero in China. Similar to what the Soviet Union did with their national space hero, Yuri Gagarin, China made the decision to not assign Yang to future space missions. However, Yang now holds the rank of Major General and is the director of the China Manned Space Engineering Office.[78]

FIRST OFFICIAL BRITON CHOSEN BY ESA & FIRST TO VISIT ISS

While he may not be the first British astronaut, Tim Peake is the first British astronaut to be sent to space by ESA. He was one of six astronauts selected by the ESA out of over 8,000 applicants in May of 2009. He is a test pilot with over 3,000 hours of flight time on more than 30 types of helicopters and fixed wing aircrafts. Peake became the first British ESA astronaut to live onboard the ISS when he launched onboard a Russian Soyuz on December 15, 2015. During his 186 days in space, Peake performed one spacewalk, participated in over 250 scientific experiments and ran the London marathon on the space station's treadmill![79]

FIRST TO SPEAK WELSH IN SPACE

Canadian astronaut Dr. Dafydd Williams is proud of his Welsh heritage. He broadcast from space in the Welsh language on his first space shuttle mission in 1998.[80] His given name is Dafydd, which is the Welsh spelling of "David."

MARRIED ASTRONAUTS & COSMONAUTS LEADS TO ASTRO-TOTS & COSMO-TOTS

Sometimes astronauts and cosmonauts marry other astronauts and cosmonauts. Some of these couples even have kids—what people have jokingly referred to as "astro-tots" and "cosmo-tots." The first woman in space, cosmonaut Valentina Tereshkova, married cosmonaut Andriyan Nikolayev and they had a daughter, Elena Andrianovna Nikolaeva-Tereshkova, born in 1964. Elena became the first person in history born to parents who had both been to space.

The third female cosmonaut in space, Yelena Kondakova, married fellow cosmonaut Valeri Ryumin in 1985 after her first flight and long-duration mission onboard the Mir space station. They had a daughter.

Yelena Serova, the fourth cosmonaut in space, married fellow cosmonaut Mark Serov, and they have one daughter. Mark retired before flying in space.[81]

Of the six women chosen in 1978 in the first NASA group of astronauts to include women (NASA Astronaut Group 8), three of them married fellow astronauts. Sally Ride and fellow Group 8 astronaut Steven Hawley married in 1982. They didn't have any children together, and divorced in 1987.

Former astronauts Rhea Seddon and Robert "Hoot" Gibson, were both in the NASA Astronaut Group 8 selection and became the first couple to meet and marry within the NASA astronaut corps. They have three children.[82]

The first mom in space, Dr. Anna Fisher—also from the NASA Astronaut Group 8 class—married a fellow astronaut, Dr. William Fisher. They had two girls, Kristin and Kara.[83]

The most recent astronauts to marry astronauts and have children include Karen Nyberg and Doug Hurley, who have a son named Jack. Astronaut Megan McArthur is married to Robert (Bob) Behnken and they also have a son.[84] Behnken and Hurley were the first astronauts to fly onboard the SpaceX Crew Dragon Demo 2. They also made history as the first American astronauts to fly from American soil since the space shuttle was retired in 2011.

FIRST MARRIED COUPLE TO FLY IN SPACE TOGETHER

Jan Davis and Mark Lee were married and flew together in space as a married couple. But they were also the last. At the time NASA had an unwritten policy not to put husbands and wives on the same mission. After Davis and Lee married during training and it was too late to replace either, NASA made a new policy banning married couples traveling to space together.[85]

NOTES

1. "Valentina Tereshkova: Whose Will the Woman Who Conquered Space Obeyed," РИА Новости, June 7, 2008, https://ria.ru/20060616/49619382.html.

2. "Soviet Union Launched First Woman Into Space Without a Toothbrush," *Moscow Times*, September 18, 2018, https://www.themoscowtimes.com/2015/09/18/soviet-union-launched-first-woman-into-space-without-a-toothbrush-a49662.

3. "Valentina Tereshkova."

4. "Valentina Tereshkova Biography," RSC ENERGIA, accessed May 24, 2021, https://www.energia.ru/english/energia/history/tereshkova/tereshkova-bio.html.

5. John Uri, "Space Station 20th – Women and the Space Station," ed. Kelli Mars, NASA, last modified March 10, 2020, https://www.nasa.gov/feature/space-station-20th-women-and-the-space-station.

6. "Biography of Svetlana Savitskaya, Russian Astronaut," Salient Women, accessed May 24, 2021, https://www.salientwomen.com/2020/05/14/biography-of-svetlana-savitskaya-russian-astronaut/.

7. Ben Evans, *Tragedy and Triumph in Orbit: the Eighties and Early Nineties* (New York: Springer, 2012), 175–176.

8. "List of Space Travelers by Nationality," Wikipedia, accessed May 24, 2021, https://en.wikipedia.org/wiki/List_of_space_travelers_by_nationality.

9. "Russia's Only Woman Cosmonaut Anna Kikina Inspires One-of-a-Kind Barbie Doll," collectSPACE.com, March 16, 2021, http://www.collectspace.com/news/news-031621a-mattel-barbie-cosmonaut-anna-kikina.html.

10. Erin Blakemore, "How Sally Ride Blazed a Trail for Women in Space," National Geographic, June 18, 2020, https://www.nationalgeographic.com/history/reference/people/sally-ride-blazed-trail-women-astronauts/.

11. Seymour Brody, "Judith Resnik Biography," Jewish Virtual Library, accessed May 24, 2021, https://www.jewishvirtuallibrary.org/judith-resnik.

12. "Judith Resnik 1949–1986," *Time*, February 10, 1986, http://content.time.com/time/subscriber/article/0,33009,960603,00.html.

13. Charlie Wood, "Kathryn Sullivan: Spacewalker and Earth Explorer," Space.com, January 13, 2020, https://www.space.com/kathryn-sullivan-bio.html.

14. Elizabeth Montoya, "Record-Breaking Highs and Lows: Meet the NASA Astronaut That Traveled to the Deepest Point on Earth," Guinness World Records, November 24, 2020, https://www.guinnessworldrecords.com/news/2020/11/record-breaking-highs-and-lows-meet-the-nasa-astronaut-that-traveled-to-the-deep-638666.

15. Nora McGreevy, "Astronaut Kathy Sullivan Becomes First Woman to Reach Deepest Part of the Ocean," *Smithsonian Magazine*, June 10, 2020, https://www.smithsonianmag.com/smart-news/kathy-sullivan-becomes-first-woman-reach-challenger-deep-180975061/.

16. Emma Doughty, "Project Juno: Pansy Seeds in Space," *The Unconventional Gardener* (blog), January 26, 2020, https://theunconventionalgardener.com/blog/project-juno-pansy-seeds-in-space/.

17. Doughty, "Project Juno."

18. "Facts about the First British Astronaut: Helen Sharman," Ørsted, accessed May 24, 2021, https://orsted.com/en/explore/space-safari/helen-sharman.

19 "Mae Jemison Had Cameo in Star Trek: The Next Generation," Peace Corps Online, January 15, 2005, http://peacecorpsonline.org/messages/messages/467/2026453.html.

20. "Biography of Roberta Lynn Bondar," Canadian Space Agency, March 8, 2018, https://www.asc-csa.gc.ca/eng/astronauts/canadian/former/bio-roberta-bondar.asp.

21. "Ellen Ochoa," Encyclopedia Britannica, May 6, 2021, https://www.britannica.com/biography/Ellen-Ochoa.

22. "JAXA Astronaut Biographies: Chiaki Mukai (M.D., Ph.D.)," Japan Aerospace Exploration Agency, last modified May 2, 2016, https://iss.jaxa.jp/en/astro/biographies/mukai/index.html.

23. Amber Karlins, "Biography: Chiaki Mukai – Astronaut," The Heroine Collective, August 31, 2015, http://www.theheroinecollective.com/chiaki-mukai-astronaut/.

24. "Europe's First Female Astronaut Appointed to French Government," *Irish Times*, June 19, 2002, https://www.irishtimes.com/news/europe-s-first-female-astronaut-appointed-to-french-government-1.1061091.

25. "Claudie Haigneré (Formerly Claudie André-Deshays)," European Space Agency, accessed May 24, 2021, https://www.esa.int/Science_Exploration/Human_and_Robotic_Exploration/Astronauts/Claudie_Haignere_formerly_Claudie_Andre-Deshays.

26. Jean-Pierre Harrison, Biography of Kaplana Chawla (Los Gatos, CA: Harrison, 2011), pdf, http://www.niscair.res.in/jinfo/sr/2012/SR%2049%285%29%20%28Book%20Review%29.pdf

27. "Profile of Liu Yang, China's First Woman Astronaut," BBC News, June 16, 2012, https://www.bbc.co.uk/news/science-environment-18471236.

28. Cathleen Lewis, "The First Mixed-Gendered Cosmonaut Candidates," National Air and Space Museum, March 30, 2017, https://airandspace.si.edu/stories/editorial/first-mixed-gendered-cosmonaut-candidates.

29. bchandler, "Alumni on Orbit," Space Camp Alumni Association, July 8, 2019, https://www.spacecampalumni.com/blog/alumni-orbit.

30. John Uri, "Space Station 20th – Women and the Space Station," ed. Kelli Mars, NASA, last modified March 10, 2020, https://www.nasa.gov/feature/space-station-20th-women-and-the-space-station.

31. Amy Sherden and Joe O'Brien, "Record-Breaking Astronaut Becomes Internet Sensation," ABC News, last modified February 23, 2016, https://www.abc.net.au/news/2015-07-08/astronaut-samantha-cristoforetti-internet-sensation/6604868.

32. "Samantha Cristoforetti," European Space Agency, accessed May 25, 2021, https://www.esa.int/Science_Exploration/Human_and_Robotic_Exploration/Astronauts/Samantha_Cristoforetti.

33. Jeanne Ryba, "STS-114 Mission Overview," NASA, last modified November 23, 2007, https://www.nasa.gov/mission_pages/shuttle/shuttlemissions/archives/sts-114.html.

34. Eileen Collins Biography," NASA, May 2006, pdf, https://www.nasa.gov/sites/default/files/atoms/files/collins_eileen.pdf.

35. "Mission Monday: Peggy Whitson Becomes the First Female ISS Commander with Expedition 16," Space Center Houston, October 5, 2020, https://spacecenter.org/mission-monday-peggy-whitson-becomes-the-first-female-iss-commander-with-expedition-16/.

36. Amy Held, "'American Space Ninja' Back on Earth after Record-Breaking Flight," NPR, September 3, 2017, https://www.npr.org/sections/thetwo-way/2017/09/03/548295156/-american-space-ninja-peggy-whitson-back-on-earth-after-record-breaking-flight.

37. Melissa Gaskill, "Celebrating Women's History Month: Most Recent Female Astronauts," ed. Michael Johnson, NASA, last modified March 3, 2021, https://www.nasa.gov/mission_pages/station/research/news/whm-recent-female-astronauts/.

38. Brian Dunbar, "Peggy A. Whitson (PH.D.) NASA Astronaut," ed. Kelli Mars, NASA, December 24, 2018, https://www.nasa.gov/astronauts/biographies/peggy-a-whitson/biography.

39. Brian Dunbar, "NASA Administration Names Whitson First NASA ISS Science Officer," NASA, September 16, 2002, https://www.nasa.gov/centers/johnson/news/releases/2002/h02-175.html.

40. Brian Dunbar, "Shannon Walker (PH.D) NASA Astronaut," ed. Melanie Whiting, NASA, May 11, 2021, https://www.nasa.gov/astronauts/biographies/shannon-walker/biography.

41. Uri, "Space Station 20th."

42. "Shannon Lucid: Enduring Qualities," NASA, accessed May 25, 2021, https://history.nasa.gov/SP-4225/nasa2/nasa2.htm.

43. Brian Dunbar, "The Incredible Shannon Lucid," NASA, accessed May 25, 2021, https://www.nasa.gov/audience/forstudents/postsecondary/features/F_The_Incredible_Shannon_Lucid_prt.htm.

44. "First All-Female Spacewalk Has Link to First US Woman to Walk in Space," collectSPACE.com, October 18, 2019, http://www.collectspace.com/news/news-101819a-first-all-female-spacewalk.html.

45. "First All-Female Spacewalk."

46. "Anna L. Fisher Biography," NASA, October 2016, PDF, https://www.nasa.gov/sites/default/files/atoms/files/fisher_anna.pdf

47. Denise Grady, "American Woman Who Shattered Space Ceiling," *New York Times*, July 23, 2012, https://www.nytimes.com/2012/07/24/science/space/sally-ride-trailblazing-astronaut-dies-at-61.html.

48. "Captain Wendy B. Lawrence Biography," United States Naval Academy Alumni Association and Foundation, accessed May 25, 2021, https://www.usna.com/events-and-programs---dga19-bio-lawrence.

49. Kyle Rempfer, "Army Astronaut Accused of Committing Crime in Space Is Cleared; Ex-Wife Charged with Making False Statements," *Army Times*, April 7, 2020, https://www.armytimes.com/news/your-army/2020/04/07/army-astronaut-accused-of-committing-crime-in-space-is-cleared-ex-wife-charged-with-making-false-statements/.

50. "The Artemis Team," NASA, accessed May 25, 2021, https://www.nasa.gov/specials/artemis-team/.

51. Zeina Saleh, "8 Things to Know About Nora Al Matrooshi, the First Female Arab Astronaut," Vogue Arabia, April 13, 2021, https://en.vogue.me/culture/nora-al-matrooshi-first-arab-female-astronaut/.

52. "Cosmonaut Kikina: Roscosmos Reinstates Rejected Female Candidate," collectSPACE.com, June 26, 2014, http://www.collectspace.com/news/news-062614a-female-cosmonaut-rejection-reversed.html.

53. Andrew Chaikin, "Neil Armstrong Didn't Lie About 'One Small Step' Moon Speech, Historian Says," Space.com, January 4, 2013, https://www.space.com/19136-neil-armstrong-moon-speech-truth.html.

54. Sarah Bruhns, "When Neil Armstrong and Edmund Hillary Took a Trip to the North Pole," Atlas Obscura, August 27, 2013, https://www.atlasobscura.com/articles/neil-armstrong-and-sir-edmund-hillarys-trip-to-the-north-pole.

55. "Buzz Aldrin, Ph.D.: Biography," NASA, January 1996, https://web.archive.org/web/20090402074622/http://www.jsc.nasa.gov/Bios/htmlbios/aldrin-b.html

56. Andrew Chaikin, *Man on the Moon: the Voyages of the Apollo Astronauts* (Great Britain: Penguin Books, 2019), 221.

57. "Michael Collins on Flying Solo during Apollo 11 Moon Landing: 'Not One Iota Lonely,'" interview by *The Guardian*, YouTube, July 16, 2019, https://youtu.be/9O572R2MwFM.

58. Mark Brown, "First Picture Drawn in Space to Appear in Cosmonauts Show in London," *The Guardian*, August 31, 2015, https://www.theguardian.com/science/2015/aug/31/first-picture-space-cosmonauts-science-museum-alexei-leonov.

59. Geoffrey Little, "John Young, Spaceman," *Air & Space Magazine*, September 2005, https://www.airspacemag.com/space/spaceman-7766826/.

60. "Bernard A. Harris, Jr., MD: Biography," NASA, January 1999, https://www.nasa.gov/sites/default/files/atoms/files/harris_bernard.pdf

61. Brian Dunbar, "NASA Deputy Administrator Frederick D. Gregory," NASA, June 22, 2007, https://www.nasa.gov/about/highlights/gregory_bio.html.

62. Carolyn Campbell, "Visit Houston Conversation with Astronauts Glover & Hopkins," Visit Houston, July 18, 2019, https://www.visithoustontexas.com/blog/post/visit-houston-conversation-with-astronauts-glover-hopkins/.

63. Brian Dunbar, "Victor J. Glover, Jr. (Commander, U.S. Navy) NASA Astronaut," ed. Melanie Whiting, NASA, last modified May 3, 2021, https://www.nasa.gov/astronauts/biographies/victor-j-glover/biography.

64. Mike Wall, "Victor Glover Becomes 1st Black Astronaut to Arrive at Space Station for Long-Term Stay," Space.com, November 17, 2020, https://www.space.com/victor-glover-first-black-crewmember-space-station.

65. Françoise Côté and Laura Neilson Bonikowsky, "Marc Garneau," The Canadian Encyclopedia, last modified February 16, 2016, https://www.thecanadianencyclopedia.ca/en/article/marc-garneau.

66. "Biography of Chris Hadfield," Canadian Space Agency, July 22, 2014, https://www.asc-csa.gc.ca/eng/astronauts/canadian/former/bio-chris-hadfield.asp.

67. Colin Burgess and Bert Vis, *Interkosmos: the Eastern Bloc's Early Space Program* (Chichester, UK: Springer International, 2016), 258.

68. "Astronauta Marcos Pontes – Vida e Realizações," MarcosPontes.com, December 2013, http://www.marcospontes.com/$SETOR/MCP/VIDA/biografia.html.

69. "Franklin Chang-Diaz: Astronaut and Rocket Scientist," PBS, November 2007, https://web.archive.org/web/20071115161244/http://www.pbs.org/kcet/wiredscience/story/80-franklin_chang_diaz_astronaut_and_rocket_scientist.html.

70. "Franklin Chang Díaz," LinkedIn, accessed May 24, 2021, https://www.linkedin.com/in/franklin-chang-d%C3%ADaz-74850750.

71. "Soviets Launch World's First Black Cosmonaut," Jet, October 9, 1980, 8, https://books.google.com/books?id=_UEDAAAAMBAJ&pg=PA8#v=onepage&q&f=false.

72. "National Hispanic Heritage Month: Carlos I. Noriega, Astronaut," Transportation History, October 6, 2020, https://transportationhistory.org/2020/10/06/national-hispanic-heritage-month-carlos-i-noriega-astronaut/.

73. "The First and Only Icelandic Astronaut Teaches a Course at RU," Reykjavik University, November 22, 2018, https://en.ru.is/news/the-first-and-only-icelandic-astronaut-teaches-a-course-at-ru.

74. "Bjarni V. Tryggvason Biography," NASA, August 2006, https://web.archive.org/web/20110525193350/http://www.jsc.nasa.gov/Bios/htmlbios/tryggvas.html.

75. "First Spanish Astronaut Carries Nation's Dreams into Space – and Takes Europe a Step into the Future," European Space Agency, October 28, 1998, https://www.esa.int/Newsroom/Press_Releases/First_Spanish_astronaut_carries_nation_s_dreams_into_space_-_and_takes_Europe_a_step_into_the_future.

76. Rym Ghazal, "The First Syrian in Space," *The National*, April 5, 2015, https://www.thenationalnews.com/arts-culture/the-first-syrian-in-space-1.38863.

77. "Dr. Franco Malerba, PH.D., Biography," March 2000, https://www.nasa.gov/sites/default/files/atoms/files/malerba.pdf

78. "Yang Liwei," Encyclopedia.com, April 15, 2021, https://www.encyclopedia.com/science/encyclopedias-almanacs-transcripts-and-maps/yang-liwei.

79. "Timothy (Tim) Peake," European Space Agency, accessed May 25, 2021, http://www.esa.int/Science_Exploration/Human_and_Robotic_Exploration/Astronauts/Timothy_Tim_Peake.

80. "Wales in Space," National Museum Wales, accessed May 25, 2021, https://museum.wales/articles/2019-10-25/Wales-in-Space.

81. Joachim Becker, "Cosmonaut Biography: Mark Serov," Space Facts, last modified April 19, 2018, http://www.spacefacts.de/bios/cosmonauts/english/serov_mark.htm.

82. "Rhea's Biography," AstronautRheaSeddon.com, February 4, 2018, https://astronautrheaseddon.com/rhea_biography/.

83. Dana Kennedy, "Fox News' Kristin Fisher Opens up about Her Childhood as an 'Astro-Tot,'" *New York Post*, May 23, 2020, https://nypost.com/2020/05/23/fox-news-kristin-fisher-on-growing-up-as-an-astro-tot/.

84. Laurel Kornfeld, "Astronaut Bob Behnken Will Be One of Two-Person Crew on Crew Dragon Demo-2 Launch," SpaceFlight Insider, May 6, 2020, https://www.spaceflightinsider.com/missions/human-spaceflight/astronaut-bob-behnken-will-be-one-of-two-person-crew-on-crew-dragon-demo-2-launch/.

85. "Love Is in the Air: Great Couples of Aerospace History," National Air and Space Museum, February 14, 2018, https://airandspace.si.edu/stories/editorial/love-air-great-couples-aerospace-history.

FALLEN, BUT NOT FORGOTTEN

It's important to remember those who gave the ultimate sacrifice in the name of space exploration. Too many have died and too many of them are unknown.

It's only been 60 years since the first human flew to space. Since then, over 30 astronauts and cosmonauts have died while training, after launch or after re-entry. Only three of those people died while in space, which is defined as above the Kármán line, which is 62 miles (100 km) above the Earth.

These are the names of those men and women who died as part of the space program.

Died During Space Flight

SOYUZ 1 DISASTER: VLADIMIR KOMAROV

The first man known to have died during a space mission launched into space on the Soyuz 1 in April 1967. After re-entering the Earth's atmosphere, the capsule's parachutes failed to open and it crashed into the ground, killing Komarov.[1]

FULL NAME: Vladimir Mikhaylovich Komarov
BORN: March 16, 1927, Moscow, Russia
DIED: April 24, 1967
SPACEFLIGHTS: Voskhod 1, Soyuz 1
▶ **DID YOU KNOW?** Komarov was the first cosmonaut to fly to space twice. His first flight was onboard Voskhod 1 with two other crewmates. They were the first to fly to space without spacesuits, due to space restrictions. One of Komarov's duties while in space was to send a radio greeting to the Tokyo Olympics, which had just begun.

X-15 CRASH: MICHAEL ADAMS

On November 15, 1967, Major Michael Adams reached 266,000 feet (81 km) above sea level in the X-15 hypersonic rocket-powered aircraft. It was his seventh X-15 flight. After reaching peak altitude, the aircraft was off its heading by 15 degrees. As it began to descend, it drifted and fell into a supersonic Mach 5 spin. Adams was able to recover from the spin at 118,000 feet (36 km), but then went into an inverted Mach 4.7 dive. The X-15 experienced forces over 15-g vertically and 8-g laterally. The craft broke up inflight, killing Adams.[2]

This flight qualified as a space flight and Adams was posthumously awarded Astronaut Wings. His name is on the Space Mirror Memorial at the Kennedy Space Center in Florida.[3]

FULL NAME: Michael James Adams
BORN: November 15, 1930, Sacramento, California
DIED: November 15, 1967
SPACEFLIGHTS: X-15 Flight 191
▶ **DID YOU KNOW?** He was a U.S. Air Force astronaut who flew the X-15 experimental spaceplane for the Air Force and NASA. He was one of the research pilots who participated in a series of NASA moon landing practice tests that lasted five months. He was the first American to die in space flight as defined by the American convention.

SOYUZ 11 DISASTER

After spending a successful and record breaking three weeks onboard the world's first space station, the Soviet Salyut 1, the crew of Soyuz 11 prepared to return home. On June 29, 1971, the Soyuz undocked from the Salyut and began its descent. After what looked like a successful re-entry, the recovery crews found all three men dead in their seats.

During separation, a faulty seal burst open on the Soyuz. The cabin depressurized and all of the air was sucked out by the vacuum of space. All three men died of suffocation. As a result, it became protocol that all cosmonauts wear pressure space suits during launch and re-entry in case of decompression, which remains in place today. These are the only three men to have died while still in space.[4]

GEORGI DOBROVOLSKY, COMMANDER
FULL NAME: Georgi Timofeyevich Dobrovolsky
BORN: June 1, 1928, Odesa, Ukraine, Soviet Union
DIED: June 30, 1971
SPACEFLIGHTS: Soyuz 11
▶ **DID YOU KNOW?** Dobrovolsky was a Soviet Air Force pilot, flight commander, deputy squadron commander, political worker and later a parachute jumper instructor.[5]

VLADISLAV VOLKOV, FLIGHT ENGINEER
FULL NAME: Vladislav Nikolayevich Volkov
BORN: Nov. 23, 1935, Moscow, Russia
DIED: June 30, 1971
SPACEFLIGHTS: Soyuz 7, Soyuz 11
▶ **DID YOU KNOW?** Volkov was the deputy leading designer involved in the construction and testing of the Vostok and Voskhod spacecraft.

VIKTOR PATSAYEV, DESIGN ENGINEER
FULL NAME: Viktor Ivanovich Patsayev
BORN: June 19, 1933, Aktyubinsk, Kazakhstan, Soviet Union (now Aqtöbe, Kazakhstan)
DIED: June 30, 1971
SPACEFLIGHTS: Soyuz 11
▶ **DID YOU KNOW?** Patsayev was the first to celebrate his birthday (38th) in space during their three weeks onboard the Salyut 1.[6]

THE SPACE SHUTTLE *CHALLENGER* DISASTER
The 25th shuttle mission was to be the first to launch a teacher into space for the Teacher in Space Project. NASA was planning to perform several experiments: deploying the second satellite in a series of

Tracking and Data Relay Satellites, flying a research tool that would observe Halley's Comet and teaching several special lessons from space. Astronaut Ron McNair was going to play a special saxophone solo that had been composed for this debut in space.

So many things went wrong for this launch. Temperatures on Florida's coast weren't supposed to fall below freezing. The flexible O-rings in the solid rocket boosters were known to fail in extreme temperatures. But warnings from engineers and safety technicians were disregarded and the launch went ahead. On the morning of January 28, 1986, at the Kennedy Space Center—after a two-hour delay to allow the ice to melt on the launchpad—*Challenger* launched at 11:38 a.m. Eastern Standard Time.

The air temperature was 15 degrees colder than it had been for any previous launch, and the O-ring seals failed as warned. *Challenger* broke apart 73 seconds into the flight and all seven of her crew members were killed.

Here are the crew of STS-51-L onboard the 10th and final flight of Space Shuttle *Challenger*:

DICK SCOBEE, COMMANDER
FULL NAME: Francis Richard Scobee
BORN: May 19, 1939, Cle Elum, Washington
DIED: January 28, 1986
SPACEFLIGHTS: STS-41-C on *Challenger*, STS-51-L on *Challenger*
▶ **DID YOU KNOW?** Scobee served in the Vietnam War as a combat aviator and was awarded the Distinguished Flying Cross, the Air Medal and other decorations. He was a test pilot before being selected as an astronaut in 1978. He was the pilot for his first shuttle mission, STS-41-C, also on Space Shuttle *Challenger*.

MIKE SMITH, PILOT
FULL NAME: Michael John Smith
BORN: April 30, 1945, Beaufort, North Carolina
DIED: January 28, 1986

SPACEFLIGHTS: STS-51-L on *Challenger*

▶ **DID YOU KNOW?** Smith was a Naval aviator and test pilot who flew 28 different types of civilian and military aircrafts and logged 4,867 hours of flying time. He also served in the Vietnam War and was awarded the Navy Distinguished Flying Cross, 3 Air Medals and several other medals and decorations. Smith had received permission to bring pieces of the famous Wright Brothers' original plane on this shuttle flight.

JUDY RESNIK, MISSION SPECIALIST

FULL NAME: Judith Arlene Resnik

BORN: April 5, 1949, Akron, Ohio

DIED: January 28, 1986

SPACEFLIGHTS: STS-41-D on *Discovery*, STS-51-L on *Challenger*

▶ **DID YOU KNOW?** Resnik played a significant part in the development of the shuttle's robotic arm and was an expert using it. She was a classical pianist, gourmet cook and the second American woman in space. She was also a Tom Selleck fan and had brought a Tom Selleck sticker to put on her shuttle locker on her previous flight.

ELLISON ONIZUKA, MISSION SPECIALIST

FULL NAME: Ellison Shoji Onizuka

BORN: June 24, 1946, Kealakekua, Kona, Hawaii

DIED: January 28, 1986

SPACEFLIGHTS: STS-51-C on *Discovery*, STS-51-L on *Challenger*

▶ **DID YOU KNOW?** Onizuka was the first Asian American and first person of Japanese origin to fly to space. He was a test pilot and flight test engineer in the U.S. Air Force.

RONALD MCNAIR, MISSION SPECIALIST

FULL NAME: Ronald Erwin McNair

BORN: October 21, 1950, Lake City, South Carolina

DIED: January 28, 1986

SPACEFLIGHTS: STS-41-B on *Challenger*, STS-51-L on *Challenger*

▶ **DID YOU KNOW?**.[7] Besides being an astronaut, McNair was a physicist who played the saxophone and had a 5th degree Karate blackbelt. McNair was the first to play a saxophone in space during STS-41-B. He also won the AAU Karate Gold Medal in 1976 with five regional Blackbelt championships.[8]

CHRISTA MCAULIFFE, PAYLOAD SPECIALIST
FULL NAME: Sharon Christa McAuliffe
BORN: September 2, 1948, Boston, Massachusetts
DIED: January 28, 1986
SPACEFLIGHTS: STS-51-L on *Challenger*
▶ **DID YOU KNOW?** McAuliffe was a social studies, history, civics and English teacher from New Hampshire. She was selected to become the first teacher to fly to space. One class she designed was called "The American Woman."[9]

GREG JARVIS, PAYLOAD SPECIALIST
FULL NAME: Gregory B Jarvis
BORN: August 24, 1944, Detroit, Michigan
DIED: January 28, 1986
SPACEFLIGHTS: STS-51-L on *Challenger*
▶ **DID YOU KNOW?** Jarvis was an engineer and worked on advanced satellite designs. He was one of two Hughes Aircraft employees selected to join the shuttle program as the first commercial customer to participate in the launch of a privately owned satellite.[10]

THE SPACE SHUTTLE *COLUMBIA* DISASTER

When Space Shuttle *Columbia* launched on January 16, 2003, a suitcase-sized piece of foam broke off from the external tank and hit *Columbia's* left wing. This damaged its heat shield. It wasn't until the day after launch that people reviewed video showing the foam breaking off and hitting the wing. This same piece of foam had been observed falling off on four previous flights throughout the years, but all four of those shuttle missions returned home without incident, and the issue became known as "foam shedding."

The engineers had major concerns about the wing damage and its safety during *Columbia's* re-entry. They made multiple requests to get pictures so they could access the damage. But NASA management felt there was no need and did nothing to inspect the damage. A week later, an email was sent to the crew of *Columbia* letting them know about the foam strike, but dismissing any concerns about re-entry.[11]

The 16-day mission went forward as normal and was dedicated to science and research with roughly 80 experiments conducted by the 7-person crew working 24 hours a day in shifts.

The mission was considered successful and productive. As the crew prepared to return to Earth on February 1, everything seemed normal—until it wasn't. As *Columbia* reentered the atmosphere, the damage from the foam caused the wing's heat shield to fail. The internal wing structure was destroyed, causing the shuttle to become unstable and break apart at a very high altitude and speed. *Columbia* disintegrated in-flight and all seven crew members were killed.[12]

The Martian hills east of the Spirit Mars Exploration Rover's landing site has been dedicated to the crew of *Columbia*. There are seven hills and each one is named after a member of the *Columbia* crew.[13]

Here are the crew of STS-107 onboard *Columbia* during its 28th and final flight:

RICK HUSBAND, COMMANDER
FULL NAME: Rick Douglas Husband
BORN: July 12, 1957, Amarillo, Texas
DIED: February 1, 2003
SPACEFLIGHTS: STS-96 on *Discovery*, STS-107 on *Columbia*
▶ **DID YOU KNOW?** Husband was a skilled pilot who logged over 3,800 hours of flight time over 40 different types of aircraft. He was the pilot who performed the first shuttle docking with the International Space Station on STS-96.[14]

WILLIE MCCOOL, PILOT
FULL NAME: William Cameron McCool

BORN: September 23, 1961, San Diego, CA
DIED: February 1, 2003
SPACEFLIGHTS: STS-107 on *Columbia*
▶ **DID YOU KNOW?** McCool's favorite song was "Imagine" by John Lennon, and he logged over 2,800 hours of flight in 24 different aircrafts with over 400 carrier landings.[15]

KALPANA CHAWLA, MISSION SPECIALIST
FULL NAME: Kalpana Chawla
BORN: July 1, 1961, Karnal, India
DIED: February 1, 2003
SPACEFLIGHTS: STS-87 on *Columbia*, STS-107 on *Columbia*
▶ **DID YOU KNOW?** Chawla was the first woman of Indian origin to go to space and is considered a national hero in India. She earned a PhD in aerospace engineering, loved hiking and backpacking and, according to her wishes, had her ashes scattered in Zion National Park.[16] Her name Kalpana means "imagination" in Hindi.

DAVID BROWN, MISSION SPECIALIST
FULL NAME: David McDowell Brown
BORN: April 16, 1956, Arlington, Virginia
DIED: February 1, 2003
SPACEFLIGHTS: STS-107 on *Columbia*
▶ **DID YOU KNOW?** Brown was a talented college gymnast who performed in the Circus Kingdom as an acrobat, 7-foot unicyclist and stilt walker. He was a Navy flight surgeon and a skilled pilot with over 2,700 flight hours. He was qualified as first pilot in the NASA supersonic T-38 spaceflight trainer jet.[17]

LAUREL CLARK, MISSION SPECIALIST
FULL NAME: Laurel Blair Clark
BORN: March 10, 1961, Ames, Iowa
DIED: February 1, 2003
SPACEFLIGHTS: STS-107 on *Columbia*

▶ **DID YOU KNOW?** Besides being an astronaut, Clark was a medical doctor, U.S. Navy Captain, Naval Flight Surgeon, Submarine Medical Officer, Diving Medical Officer, Undersea Medical Officer and Radiation Health Officer. She also helped create a treadmill for the International Space Station.[18]

MICHAEL ANDERSON, MISSION SPECIALIST
FULL NAME: Michael Phillip Anderson
BORN: December 25, 1959, Plattsburgh, New York
DIED: February 1, 2003
SPACEFLIGHTS: STS-89 on *Endeavour*, STS-107 on *Columbia*
▶ **DID YOU KNOW?** Anderson had a Master's degree in physics, was an Air Force pilot and flight instructor, and was the payload commander and lieutenant colonel in charge of science experiments on *Columbia*. He drove a Porsche but never got a speeding ticket.[19] He also sang tenor in his church choir.[20]

ILAN RAMON, PAYLOAD SPECIALIST
FULL NAME: Ilan Ramon
BORN: June 20, 1954, Ramat Gan, Israel
DIED: February 1, 2003
SPACEFLIGHTS: STS-107 on *Columbia*
▶ **DID YOU KNOW?** Ramon was the first and only Israeli astronaut, and the first to request kosher food and mark the Sabbath as a way of representing all Jews and all Israelis.[21] When Israel's first Moon mission launched in 2019, among the items it carried to the Moon was a time capsule with a photo of Ramon.[22]

VIRGIN GALACTIC SPACEPLANE CRASH: MICHAEL ALSBURY
During a test flight of Virgin Galactic's SpaceShipTwo VSS *Enterprise* on October 31, 2014, the spaceplane broke apart in-flight. The pilot survived, but the co-pilot Michael Alsbury didn't. His name was added to the Space Mirror at Kennedy Space Center on January 25, 2020.

FULL NAME: Michael Tyler Alsbury
BORN: Mar. 19, 1975, Santa Clara, California
DIED: October 31, 2014
SPACEFLIGHTS: None
▶ **DID YOU KNOW?** Alsbury was a Virgin Galactic Commercial Astronaut and had flown eight times aboard SpaceShipTwo before his death on what was his 9th flight. He was a highly experienced and respected pilot who had flown over 1,800 hours—1,600 of them as a test pilot and engineer with the company Scaled Composites, who built SpaceShipTwo.[23]

Died During Training or Testing

ALTITUDE CHAMBER FIRE: VALENTIN BONDARENKO

On March 23, 1961, just shy of three weeks before the first human flew to space, 24-year-old cosmonaut Valentin Bondarenko was inside an altitude chamber filled with pure oxygen at pressure. No one had yet flown to space and scientists were still studying how men would handle long periods of isolation.

After spending several days in the chamber, a small fire started inside it. The fire quickly became uncontrollable due to the highly combustible pure oxygen inside the chamber. Because the chamber was pressurized, the doors couldn't be opened immediately and it took several minutes to get him out. Bondarenko had been severely burned and died hours later.[24]

FULL NAME: Valentin Vasiliyevich Bondarenko
BORN: February 16, 1937, Kharkov, Ukraine, Soviet Union
DIED: March 23, 1961
SPACEFLIGHTS: None
▶ **DID YOU KNOW?** Valentin Bondarenko was an air force pilot and one of the first 20 men selected to become cosmonauts. His death was originally covered up by the Soviet Union and was only revealed in 1980.

FOG, GEESE AND A T-38 JET: TED FREEMAN

A year after being selected as an astronaut, Ted Freeman was flying to Houston from the McDonnell training facilities in St. Louis, Missouri, on October 31, 1964. There were reports of fog and geese, and he collided with a goose. Freeman's engine flamed out and he attempted to land on the runway, but was coming in too short and didn't want to crash into military housing. He banked away and bailed out, but not in time. His parachute didn't open and he died on impact.

FULL NAME: Theodore Cordy Freeman
BORN: February 18, 1930, Haverford, Pennsylvania
DIED: October 31, 1964
SPACEFLIGHTS: None
▶ **DID YOU KNOW?** When he was younger, Freeman worked part-time at the airport refueling planes and working on them. He used that money to pay for flying lessons, and earned his private pilot's license at the age of 16. Later, Freeman became a test pilot and earned a Master's degree in aeronautical engineering. Freeman was assigned to help develop boosters at NASA while waiting for his mission assignment.

NASA T-38 JET CRASH

On February 28, 1966, four astronauts were flying in two separate T-38 jets headed for St. Louis, Missouri. Tom Stafford and Gene Cernan were in one jet. Charlie Bassett and Elliot See were in the other. Poor visibility due to bad weather caused Bassett and See to misjudge their approach and they crashed into the McDonnell Aircraft Building 101 where the Gemini spacecraft was built.

During the memorial service, astronauts Jim McDivitt and Jim Lovell and civilian pilot Jerrie Cobb honored Bassett by flying an aerial salute in what's known as the missing man formation. The jets perform a flyby in formation designed for four planes, but the fourth spot is empty to symbolize the pilot's absence. The same was done for See by fellow astronauts Buzz Aldrin, Bill Anders and Walter Cunningham.[25]

CHARLIE BASSETT
FULL NAME: Charles Arthur Bassett II
BORN: December 30, 1931, Dayton, Ohio
DIED: February 28, 1966
SPACEFLIGHTS: Assigned to Gemini 9
▶ **DID YOU KNOW?** Bassett made his first flight at age 16 and worked odd jobs to pay for flying lessons. He earned his private pilot license at 17. Bassett had a Bachelor's degree in electrical engineering and was an experimental test pilot with over 3,600 hours of flying time.

ELLIOTT SEE
FULL NAME: Elliot McKay See Jr.
BORN: July 23, 1927, Dallas, Texas
DIED: February 28, 1966
SPACEFLIGHTS: Assigned to Gemini 9
▶ **DID YOU KNOW?** See was tasked with determining whether the crewed lunar landing should occur in direct sunlight or in earthshine (using light reflected from the Earth). To simulate earthshine, he flew helicopters and planes over lava flows in California wearing dark welding goggles. His research led to astronauts landing during the day in the sunlight, but not directly under the sun.[26]

NASA T-38 CRASH: C.C. WILLIAMS
While flying from Cape Canaveral back to Houston in October 1967, C.C. Williams wanted to stop in Mobile, Alabama, to visit his father, who was dying of cancer. While flying over Tallahassee, Florida, his T-38 jet had a mechanical failure that caused it to roll to the left suddenly and then dive straight down. As trained, Williams radioed a mayday and ejected from the aircraft, but the craft was traveling too fast and too low and his parachute didn't open properly. Williams was killed, and never got to fly the Lunar Module he had been assigned to work on while at NASA.[27]

FULL NAME: Clifton Curtis Williams, Jr.

BORN: Sept. 26, 1932, Mobile, Alabama

DIED: October 5, 1967

SPACEFLIGHTS: Backup pilot for Gemini 10

▶ **DID YOU KNOW?** Williams should have been the 4th person to walk on the Moon as the Lunar Module pilot on Apollo 12. After his death, Alan Bean took his place. In a missing man formation-style salute, the three-man crew of Apollo 11 included four stars on their mission patch. One star for each of the three astronauts onboard and one for Williams, as suggested by Bean. Bean also placed William's naval aviator wings and silver astronaut pin on the surface of the Moon as a tribute.[28]

DROWNING ACCIDENT DURING TRAINING: SERGEI VOZOVIKOV

Three cosmonauts were participating in water survival and recovery training in the Black Sea on July 11, 1993. Cosmonaut Sergei Vozovikov became entangled in an old fishing net and drowned.

FULL NAME: Sergei Yuriyevich Vozovikov

BORN: April 17, 1958, Alma-Ata, Kasakh, Soviet Union

DIED: July 11, 1993

SPACEFLIGHTS: None

▶ **DID YOU KNOW?** Vozovikov had been selected as a military cosmonaut candidate in May 1990. He had completed his cosmonaut training and had been assigned to the Soyuz-TM missions to Mir.[29]

PLANE CRASH: YURI GAGARIN

On April 12, 1961, Yuri Gagarin made history when he became the first human in space. After his first and only flight to space, Gagarin served as a deputy to the Soviet Union and spent time at the cosmonaut facility working on reusable spacecraft designs. Almost 7 years after his historic flight, Gagarin died in a plane crash during a routine training flight on March 27, 1968

FULL NAME: Yuri Alekseyevich Gagarin
BORN: March 9, 1934, Klushino, Smolensk Oblast, Soviet Union
DIED: March 27, 1968
SPACEFLIGHTS: Vostok 1
▶ **DID YOU KNOW?** Gagarin played ice hockey as the goalkeeper in his youth. As a tribute, the Kontinental Hockey League named their championship trophy the Gagarin Cup in 2008.[30]

APOLLO 1 FIRE

During a test run on the launchpad for Apollo 1 in January 1967, a stray spark led to an uncontrollable fire in the cabin. Since the cabin was filled with pure oxygen, the fire burned quickly. The crew were unable to get the hatch open and all three astronauts died.[31]

GUS GRISSOM

FULL NAME: Virgil Ivan Grissom
BORN: October 5, 1929, Seattle Washington
DIED: January 27, 1967
SPACEFLIGHTS: Mercury-Redstone 4, Gemini 3, Apollo 1(not flown)
▶ **DID YOU KNOW?** Grissom was the second American to fly in space and the first of the Mercury Seven to make a second trip to space on the Gemini 3 mission. Grissom served in both World War II and the Korean War. Before he was selected as an astronaut, Grissom was a mechanical engineer and test pilot.

ED WHITE

FULL NAME: Edward Higgins White, II
BORN: November 14, 1930, San Antonio, Texas
DIED: January 27, 1967
SPACEFLIGHTS: Gemini 4, Apollo 1 (not flown)
▶ **DID YOU KNOW?** White was the first American to walk in space. He was one of the nine men selected in the second group

of NASA astronauts. White was considered to be the most physically fit of all the astronauts in the corp.[32]

ROGER CHAFFEE
FULL NAME: Roger Bruce Chaffee
BORN: February 15, 1935, Grand Rapids, Michigan
DIED: January 27, 1967
SPACEFLIGHTS: Apollo 1 (not flown)
▶ **DID YOU KNOW?** Chaffee was the youngest American astronaut to be selected for amission at the time. He served as CAPCOM for the Gemini 3 and 4 missions. He was a gun collector and would even make his own rifles.[33]

CAR ACCIDENT: EDWARD GIVENS

Astronaut Edward Givens had been assigned to the Apollo program and was destined to be a household name, but he died before he had the chance on June 6, 1967. He was an accomplished and skilled test pilot. Before he had been assigned to a specific Apollo mission, Givens was in a car accident on a rainy night. He died from injuries he sustained in the accident. Before his death, Givens had served as support crew for Apollo 1 and Apollo 7.[34]

FULL NAME: Edward Galen Givens Jr.
BORN: January 5, 1930, Quanah, Texas
DIED: June 6, 1967
SPACEFLIGHTS: None
▶ **DID YOU KNOW?** Givens was one of the astronaut candidate finalists for Project Mercury before being selected seven years later in the fifth group of astronauts.

FLIGHT ACCIDENT: ROBERT LAWRENCE JR.

In December 1967 during a routine training flight, Lawrence was training a pilot on the steep-descent glide technique. The trainee didn't flare (make a final adjustment for a safe landing) in time, and the plane hit the ground hard. The landing gear also failed, and the plane caught fire and rolled as it bounced and skidded on the runway. The trainee in the front seat was able to eject from the plane and survived. Lawrence—who was in the backseat—also ejected, but not in time to avoid hitting the front seat. He ejected sideways and he was killed instantly.[35]

FULL NAME: Robert Henry Lawrence Jr.
BORN: October 2, 1935, Chicago, Illinois
DIED: December 8, 1967
SPACEFLIGHTS: None
▶ **DID YOU KNOW?** Lawrence was selected by the U.S. Air Force as an astronaut in the Air Force's Manned Orbital Laboratory space program. He was America's first black astronaut and most likely would have flown on the Space Shuttle had he not been killed. He was a senior Air Force pilot whose experience and tests contributed significantly to the development of the Space Shuttle. He also had his Ph.D. in physical chemistry.[36]

DIED OF ILLNESS: PAVEL BELYAYEV

Five years after cosmonaut Pavel Belyayev flew on the Voskhod 2, he died in 1970 from peritonitis that developed after a stomach ulcer operation.[37] His name is one of the 14 included on the name plaque next to the Fallen Astronaut figurine on the Moon.

FULL NAME: Pavel Ivanovich Belyayev
BORN: June 26, 1925, Chelishchevo, now Babushkinsky District, Russia
DIED: January 10 1970
SPACEFLIGHTS: Voskhod 2
▶ **DID YOU KNOW?** Belyayev is believed to have been the first cosmonaut to manually fly a Soviet spacecraft after the autopilot landing system failed. He and Alexei Leonov made it back to Earth safely, but 300 miles (483 km) off course. They spent a cold night in their capsule with hungry wolves lurking outside while they waited for the rescue team. After

this experience, a small, lightweight pistol was included in the emergency survival kits for cosmonauts.[38]

STUNT PLANE ACCIDENT: STEPHEN THORNE

In May 1986, astronaut candidate Stephen Thorne was a passenger in a small stunt plane with pilot and NASA flight control engineer James Ryan Simons. The stunt plane was in the middle of performing several maneuvers near Santa Fe, Texas. The small plane went into an inverted tailspin and crashed, killing both Thorne and Simons.[39] Thorne had only been an astronaut candidate for 11 months when he died in the accident.

FULL NAME: Stephen Douglas Thorne
BORN: February 11, 1953, Frankfurt, West Germany
DIED: May 24, 1986
SPACEFLIGHTS: None
▶ **DID YOU KNOW?** Thorne first applied to be part of the Astronaut Group 10 in 1984, but was unsuccessful. A year later in June of 1985 he was accepted as an astronaut candidate as part of Astronaut Group 11.[40]

VINTAGE WORLD WAR II PLANE CRASH: S. DAVID GRIGGS

While practicing for an airshow later that day, astronaut S. David Griggs, was performing aerobatic maneuvers in a vintage World War II plane and lost control and crashed near Earle, Arkansas in June 1989. Griggs had been assigned as prime crew for an upcoming space shuttle mission that November.[41] Astronaut John Blaha flew in his place as the pilot of STS-33.

FULL NAME: Stanley David Griggs
BORN: September 7, 1939, Portland, Oregon
DIED: June 17, 1989
SPACEFLIGHTS: STS-5-D on *Discovery*
▶ **DID YOU KNOW?** During his first and only spaceflight, he performed the first unscheduled EVA (spacewalk) to prepare for a satellite rescue attempt.[42]

COMMERCIAL AIRPLANE CRASH: SONNY CARTER

This regularly scheduled commuter flight from Atlanta, Georgia, to Brunswick, Georgia, was carrying 23 people. The plane was within eyesight from the Jacksonville airport in Atlanta when it suddenly rolled to the left and dove nose down into the ground two miles from the runway. The accident was caused by a propeller system malfunction, which caused the sudden roll. All 23 people onboard died, including astronaut Sonny Carter, who was traveling for NASA on April 4, 1991.

FULL NAME: Manley Lanier Carter Jr.
BORN: August, 15, 1947, Macon, Georgia
DIED: April 4, 1991
SPACEFLIGHTS: STS-33 on *Discovery*
▶ **DID YOU KNOW?** Carter was a professional soccer player while attending medical school. He was also a chemist, flight surgeon, fighter pilot and test pilot. NASA named The Sonny Carter Training Facility in his honor. This is where NASA's Neutral Buoyancy Laboratory (NBL) is located, where astronauts train underwater for spacewalks.[43]

PRIVATE PLANE ACCIDENT: PATRICIA HILLIARD ROBERTSON

While performing touch-and-go landings in a two-seat experimental airplane in Manvel, Texas, passenger Patricia Hilliard Robertson and the pilot crashed. She and the pilot were critically injured and had second and third-degree burns over 90 percent of their bodies. They both died two days later on May 24, 2001.[44]

FULL NAME: Patricia Consolatrix Hilliard Robertson
BORN: March 12, 1963, Indiana, Pennsylvania
DIED: May 24, 2001
SPACEFLIGHTS: None
▶ **DID YOU KNOW?** Before being selected as an astronaut, Robertson worked as a physician at the Flight Medicine Clinic at the Johnson Space Center where she treated astronauts and

their families. Besides being a physician and astronaut, she was a multi-engine rated flight instructor and aerobatic pilot. Robertson was selected as an astronaut in June 1998. She was scheduled to fly on her first mission as an astronaut to the International Space Station in 2002, but died before she was able to fly into space.

NOTES

1. Robert Krulwich, "A Cosmonaut's Fiery Death Retold," NPR, May 3, 2011, https://www.npr.org/sections/krulwich/2011/05/03/135919389/a-cosmonauts-fiery-death-retold.

2. "Michael J. Adams – Biography," NASA, accessed May 28, 2021, https://history.nasa.gov/x15/adams.html.

3. "The Crash of the X-15A-3," Check-Six.com, last modified August 25, 2018, http://www.check-six.com/Crash_Sites/X-15A_crash_site.htm.

4. Jake Parks, "How Many People Have Died in Outer Space?" *Discover Magazine*, October 8, 2019, https://www.discovermagazine.com/the-sciences/how-many-people-have-died-in-outer-space.

5. Colin Burgess, Kate Doolan, and Bert Vis, *Fallen Astronauts: Heroes Who Died Reaching for the Moon* (Lincoln, NE: University of Nebraska, 2016), 184, https://books.google.com/books?id=iJ8WwRBNgkOC.

6. Burgess, Doolan, and Vis, *Fallen Astronauts*, 186.

7. Kurt Heisig, "Sax in Space," Kurt Heisig Music, accessed May 28, 2021, https://www.kurtheisigmusic.com/sax-in-space/.

8. John Stevenson, "BHM Remembers: Dr Ronald McNair," Black History Month 2021, October 21, 2015, https://www.blackhistorymonth.org.uk/article/section/science-and-medicine/bhm-remembers-robert-mcnair/.

9. Colin Burgess, *Teacher in Space: Christa McAuliffe and the Challenger Legacy* (Lincoln, NE: University of Nebraska Press, 2020), 16.

10. Jack Fisher, "The Hughes Astronauts," *Our Space Heritage 1960–2000* (blog), June 4, 2013, https://www.hughesscgheritage.com/hc-payload-specialists-signal-hughes-communications-newsletter/4/.

11. Columbia Accident Investigation Board, "Decision Making at NASA" (Report vol. I, chap. 6), August 2003, https://web.archive.org/web/20110516132723/http://caib.nasa.gov/news/report/pdf/vol1/chapters/chapter6.pdf.

12. Elizabeth Howell, "Columbia Disaster: What Happened and What NASA Learned," Space.com, February 1, 2019, https://www.space.com/19436-columbia-disaster.html.

13. Brian Dunbar, "NASA Dedicates Mars Landmarks to Columbia Crew," ed. Jim Wilson, NASA, November 25, 2007, https://www.nasa.gov/home/hqnews/2004/feb/HQ_04048_columbia_landmarks.html.

14. "Astronaut Bio: Rick D. Husband," NASA, February 2003, https://history.nasa.gov/columbia/Troxell/Columbia%20Web%20Site/Biographies/Crew%20Profile%20Information/Crew%20Biographies/ASTRON~2.HTM.

15. "Astronaut Bio: William McCool," NASA, February 2003, https://history.nasa.gov/columbia/Troxell/Columbia%20Web%20Site/Biographies/Crew%20Profile%20Information/Crew%20Biographies/McCool%20Astronaut%20Bio%20Data.htm.

16. "Kalpana Chawla, Ph.D.: Biography," NASA, May 2004, https://www.nasa.gov/sites/default/files/atoms/files/chawla_kalpana.pdf

17. "Astronaut Bio: David Brown," NASA, February 2003, https://history.nasa.gov/columbia/Troxell/Columbia%20Web%20Site/Biographies/Crew%20Profile%20Information/Crew%20Biographies/Brown%20Astronaut%20Bio%20Data.htm.

18. "Laurel Clark," Wisconsin Women Making History (Women in Wisconsin, August 26, 2020), https://womeninwisconsin.org/profile/laurel-clark/.

19. Jonathan Martin and David Postman, "Michael Anderson: From Humble Roots to One of America's 'Humble Heroes,'" Seattle Times, February 9, 2003, https://archive.seattletimes.com/archive/?date=20030209&slug=anderson09m.

20. "Bio: Shuttle Columbia Payload Commander Lt. Col. Michael Anderson," Fox News, last modified January 13, 2015, https://www.foxnews.com/story/bio-shuttle-columbia-payload-commander-lt-col-michael-anderson.

21. Israel Hanukoglu, "Ilan Ramon – The First Israeli Astronaut," Israel Science and Technology Directory, accessed May 28, 2021, https://www.science.co.il/Ilan-Ramon/.

22. Hanneke Weitering, "Israeli Moon Lander Team Draws Inspiration from Country's 1st Astronaut Ilan Ramon," Space.com, February 21, 2019, https://www.space.com/spacils-moon-mission-honors-first-israeli-astronaut.html.

23. "Fallen SpaceShipTwo Pilot's Name Added to Space Mirror Memorial," collectSPACE.com, January 25, 2020, http://www.collectspace.com/news/news-012520a-space-mirror-memorial-alsbury-commercial-astronaut.html.

24. Burgess, Doolan, and Vis, "Fallen Astronauts," 163–165.

25. "Astronauts Are Bid Farewell in Texas," The Record, March 3, 1966, https://www.newspapers.com/clip/29300913/the-record/.

26. "Moon Explorers Face Light Problems," Indianapolis Star, August 23, 1964, p. 18, https://www.newspapers.com/image/?clipping_id=35559995.

27. "Clifton Williams: Astronaut Killed in Jet Mishap," Mt. Vernon Register-News, October 6, 1967, https://www.newspapers.com/clip/29382477/mt-vernon-register-news/.

28. Brian Dunbar, "50 Years Ago, on the Way to the Moon...," ed. Mark Garcia, NASA, October 5, 2017, https://www.nasa.gov/feature/50-years-ago-on-the-way-to-the-moon-astronaut-clifton-c-cc-williams.

29. David Shayler, Disasters and Accidents in Manned Spaceflight (London: Springer, 2000).

30. Michael Long, "UFA Sports to Market Kontinental Hockey League," Sports Pro, May 19, 2019, https://www.sportspromedia.com/news/ufa_sports_to_market_kontinental_hockey_league.

31. Parks, "How Many People."

32. Mary C. White, "Detailed Biographies of Apollo I Crew – Ed White," NASA, last modified August 4, 2006, https://history.nasa.gov/Apollo204/zorn/white.htm.

33. Mary C. White, "Detailed Biographies of Apollo I Crew – Roger Chaffee," NASA, August 4, 2006, https://history.nasa.gov/Apollo204/zorn/chaffee.htm.

34. Burgess, Doolan, and Vis, "Fallen Astronauts," 190–216.

35. "First Negro Astronaut Killed in Plane Crash," Eugene Register-Guard, December 9, 1967, p. 1, https://news.google.com/newspapers?id=7uVVAAAAIBAJ&pg=4979%2C1973906.

36. "Scholarships," Bradley University, accessed May 28, 2021, https://www.bradley.edu/academic/departments/chm/scholarships/.

37. Burgess, Doolan, and Vis, "Fallen Astronauts," 179–181.

38. Mary Roach, Packing for Mars: The Curious Science of Life in the Void (New York: WW Norton, 2011), 70.

39. "Two Killed in Plane Crash," Associated Press, May 24, 1986, https://apnews.com/article/bfba3daa911ccddc99da7899f575cc74.

40. "Stephen Douglas Thorne Astronaut Biography," SpaceFacts.de, accessed October 17, 2021, http://www.spacefacts.de/bios/astronauts/english/thorne_stephen.htm.

41. "Astronaut Killed in Air Crash May Have Broken NASA Rule," *Los Angeles Times*, June 19, 1989, https://www.latimes.com/archives/la-xpm-1989-06-19-mn-1869-story.html.

42. "Astronaut S. David Griggs Biography," NASA, June 1989, https://www.nasa.gov/sites/default/files/atoms/files/griggs_s_david.pdf.

43. "Astronaut Manley Lanier 'Sonny' Carter, Jr. Biography," NASA, April 1991. https://www.nasa.gov/sites/default/files/atoms/files/carter_manley_0.pdf

44. "NASA Astronaut Killed in Plane Crash," AirSafe.com, September 27, 2014, http://www.airsafe.com/events/celebs/nasa.htm.

MISSION PATCHES, BADGES & PINS

When the space program started, the people first sent to space were most often pilots with military backgrounds. In the military, it's tradition that pilots have shoulder patches on their flight suits. This tradition carried over to the men and women who have flown to space.[1]

If you pay close attention, you'll notice that astronauts and cosmonauts today have a variety of mission patches because there are different types.[2] Astronauts are also awarded badges and pins for several reasons.

TYPES OF MISSION PATCHES

• **MISSION PATCH OR CREW PATCH** – They are designed for each space mission and are personalized for each mission. The mission patches often feature the crew members' names. While the Chinese space program has mission patches, they don't include the crews' names.

• **PAYLOAD PATCH** – These were designed for a particular payload carried onboard the space shuttle missions.

• **ISS EXPEDITION PATCH** – Each expedition to the International Space Station has its own patch, and the crew assigned to that expedition wears that patch. They also wear a separate patch for the spacecraft—Soyuz, Shuttle, or Crew Dragon—that takes them to the station.

• **PROJECT PATCH** – Experiments, programs or projects sometimes have their own patches. Spaceflight programs like Apollo, Space Shuttle and more recently, SpaceX, have program logos.

- **AGENCY PATCH** – Each space agency has a patch with its logo. NASA has its round red, white and blue logo, ESA has a "thumbprint" or "flags" patch, JAXA has a star and the Roscosmos patch has a red arrow surrounded by an orbit both of which are inside a white circle.

- **PERSONAL PATCH** – These patches aren't normally shown to the public, but sometimes personal patches are created for a specific astronaut to use on one or more space missions.

- **GROUP PATCH** – Astronauts and cosmonauts are selected in groups and sometimes give themselves nicknames. Patches are sometimes designed that feature the logo for that group of astronauts or cosmonauts, which usually include the year they were selected.

- **UNMANNED & COMMEMORATIVE PATCHES** – Not everyone considers these "true" space patches, but these are patches that have been created for rockets, satellites, spaceships, the Moon, planets or stars.

THE FIRST SPACE PATCH
Cosmonaut Valentina Tereshkova was the first astronaut to wear a space patch when she flew to space in 1963 onboard the Vostok 6. The patch wasn't visible to the public because it had been sewn into her inner suit and worn under her bright orange coveralls.[3]

THE FIRST SPACE PATCH SEEN IN PUBLIC
Cosmonaut Alexei Leonov, the first man to perform a spacewalk, wore the arrowhead CCCP logo patch on the outside of his Berkut spacesuit. This was the first time a space patch was seen in public.[4]

NASA PATCHES
It wasn't until the Gemini 5 flight that mission patches become a tradition for NASA. The first mission patch featured a covered wagon

The very first mission patch, created for Gemini 5.
Image courtesy of NASA.

with the last names, Cooper and Conrad to represent the crew flying onboard. Since then, all manned space missions have had patches. There have been 8 designs for Gemini, 12 for Apollo, 3 for Skylab, 1 for the Apollo-Soyuz Test Project, 135 for the Space Shuttle program and 3 for Space X (as of April 2021).

ESA PATCHES

The first patch to represent a European flying into space was in 1978 when cosmonaut Vladimír Remek flew onboard Soyuz 28. He was the first Czech in space and the first cosmonaut from a country other than Russia or the United States. You can find all of the flight and mission patches worn by European astronauts here: http://www. esa.int/About_Us/ESA_history/European_human_spaceflight_ patches

UNITED STATES ASTRONAUT BADGES

Astronauts who fly for NASA are awarded a special astronaut badge. There are two different types of astronaut badges: one for astronauts who have served in the military and one for civilian astronauts.

MILITARY BADGES

The U.S. Air Force, Army, Navy, Marine Corps and Coast Guard each have slightly different versions of their astronaut badges. These are the least-awarded badge of qualification in the military.[5] To earn it, an officer must complete all required training and participate in a space flight more than 50 miles (80 km) above Earth.

CIVILIAN BADGES

Astronauts who are involved in U.S. space missions but who are not in the military are awarded badges by NASA.

NASA SPACE SHUTTLE PAYLOAD SPECIALIST BADGES

A special badge was made for astronauts who served as payload specialists on the space shuttles. Payload specialists were often individuals chosen by the research community, were non-NASA astronauts chosen by partner nations or were representatives of the U.S. legislature.

COMMERCIAL ASTRONAUT WINGS

Commercial astronauts who fly in privately-funded spacecraft can be granted commercial astronaut wings by the U.S. FAA (Federal Aviation Administration) if they fly more than 50 miles (80 km) above Earth.[6]

NASA ASTRONAUT PINS

Each NASA astronaut who successfully completed astronaut training is awarded a silver pin. Astronauts who fly to space can be awarded a gold pin. But they have to pay out of pocket for those, around $400![7] Astronaut pins have been a NASA tradition since the Mercury 7 astronauts.[8]

NON-NASA ASTRONAUT PINS

Cosmonauts who have made a spaceflight have been presented with a badge ever since Yuri Gagarin's first flight. The British Interplanetary Society has given UK astronauts a rocket-shaped pin.

UNIVERSAL ASTRONAUT INSIGNIA

In April 2021, the Association of Space Explorers (ASE) created a pin for anyone who has flown into space. The Universal Astronaut Insignia has two versions. One for those who have made a suborbital spaceflight and returned to Earth and one version for those who have entered Earth's orbit or flew beyond.

SILVER PIN ON THE MOON

Astronaut Clifton Williams was killed in a plane crash before he was to fly to the Moon. Alan Bean was his replacement, and flew to the Moon on Apollo 12 in 1969.[9] To honor Williams, Bean took Williams's naval aviator wings and placed them on the surface of the Moon.[10] Bean took his own silver astronaut pin and threw it into a crater on the Moon, where it will stay for millions of years.[11]

ONE-OF-A-KIND PIN

Astronaut Deke Slayton was one of the original Mercury 7 astronauts, but was grounded for medical reasons and never flew during the Mercury or Gemini programs. It was thought that he would never make it to space.

After being medically grounded, Slayton was reassigned to work in senior management at NASA and was responsible for crew assignments.

Selection of Astronaut & Space Explorer Pins and Badges. Image montage credit: collectSpace.com (modified by the author for this edition.)

Left column: Badges awarded to astronauts in the U.S. military who fly to space. Different designs depending on the branch of military. Images courtesy of the Smithsonian.

Center column: Top: NASA astronaut pins–silver when selected, gold when flown. Center: Commercial Astronaut Wings awarded by the U.S. FAA. Bottom: The first badge awarded to Yuri Gagarin in April 1961, the Pilot-Cosmonaut of the USSR Badge. Images courtesy of NASA, FAA and RIA Novoski/Wikipedia.

Right column: Pins awarded by Virgin Galactic, Blue Origin and SpaceX to those who fly on their spacecrafts. Image credits: Virgin Galactic and Mark Bonner, Blue Origin and John Kraus, Inspiration4.

To honor Slayton for his work at NASA, the crew of Apollo 1 requested that a special astronaut pin be made. Since he was largely responsible for astronauts being assigned to fly to space, the crew felt like a special gold pin would show their appreciation.

A gold pin—much like the one given to astronauts who fly to space—was given to Slayton. But a diamond was added to make it unique. It was supposed to fly to space with the crew of Apollo 1, but the crew died in a tragic fire during a routine system check. The pin eventually made it to the Moon on Apollo 11 and was brought back and worn by Slayton. The pin was presented to Slayton by the wives of the Apollo 1 crewmembers.[12]

Slayton was eventually assigned to fly 16 years after being selected as an astronaut. He received the normal gold astronaut pin, but continued to wear his one-of-a-kind pin.

NOTES

1. Eric Wollbrett, "Les Badges Spatiaux," Space Badges, accessed May 30, 2021, https://space-badges.pagesperso-orange.fr/.

2. Travis K. Kircher, "Mission Patches: More Than Just a Merit Badge," collectSPACE, accessed May 30, 2021, http://www.collectspace.com/resources/patches_astronauts.html.

3. "'8 Days or Bust' +50 Years: Gemini 5 Made History with First Crew Mission Patch," collectSPACE.com, August 24, 2015, http://www.collectspace.com/news/news-082415a-gemini5-50th-8daysorbust.html.

4. "Salyut – CCCP Zvezda-Rocket (1965)," SpacePatches, accessed May 30, 2021, http://www.spacepatches.nl/salyut/cccp.html.

5. Alex McVeigh, "First Tomb Badge Recipient Laid to Rest," Army.mil, February 11, 2011, https://www.army.mil/article/16782/first_tomb_badge_recipient_laid_to_rest.

6. "U.S. Navy Naval Astronaut Badge," US Military Uniforms & Insignia Manufacturer, accessed May 30, 2021, http://www.uniforms-4u.com/p-astronaut-breast-insignia-3169.aspx.

7. Mike Mullane, *Riding Rockets: the Outrageous Tales of a Space Shuttle Astronaut* (New-York: Scribner, 2006), 192.

8. Robert Z. Pearlman, "'Pin'-Nacle Achievement: The Story Behind NASA's Astronaut Pin," Space.com, January 6, 2020, https://www.space.com/nasa-astronaut-pin-history.html.

9. Pearlman, "'Pin'-Nacle Achievement."

10. Brian Dunbar, "50 Years Ago, on the Way to the Moon...," ed. Mark Garcia, NASA, October 5, 2017, https://www.nasa.gov/feature/50-years-ago-on-the-way-to-the-moon-astronaut-clifton-c-cc-williams.

11. Alan Bean, "Lone Star," Alan Bean Gallery, accessed May 30, 2021, http://www.alanbeangallery.com/lonestar-story.html.

12. Pearlman, "'Pin'-Nacle Achievement."

HOW MUCH THINGS COST

PUTTING MAN ON THE MOON: $175 BILLION

The Apollo Program cost almost $25 billion in 1969. That's about $176 billion in 2020 money.[1] The money was spent on rockets, spacecrafts, computers, ground control and paying the more than 400,000 people involved in making President John F. Kennedy's goal to put man on the Moon a success.[2] Cost per seat works out to be $390 million in 2020 dollars.[3]

APOLLO SPACESUIT: $100,000

The suits moonwalkers wore had to protect them from extreme temperatures, and moon dust that was as sharp as glass. They also had to provide astronauts with the air and everything else they needed to keep them alive.

THE LUNAR MODULE: $388 MILLION

No one had ever built anything like this before but it had to work, or else people died in space or on the Moon. This spacecraft had to land safely on the Moon, and get astronauts back to the Command Module spacecraft that was orbiting the Moon so it could take them home.[4]

MIR SPACE STATION: $4.2 BILLION

From 1986 to 2011, the Mir space station hosted 125 cosmonauts and astronauts from 12 different nations. It supported 17 space expeditions, including 28 long-term crew. It served as a floating laboratory for 23,000 scientific and medical experiments. It's estimated to have cost $4.2 billion during its 15 years in space.[5]

SKYLAB SPACE STATION: $2.2 BILLION

Over the course of its nine-year existence from 1966 to 1974, America's first space station hosted nine astronauts from three separate missions who spent a total of 510 days onboard Skylab. Skylab cost $2.2 billion in 1974 money. In 2020, that works out to be $11.5 billion.[6]

SPACE SHUTTLE PROGRAM

Throughout the life of the program, it's estimated to have cost almost $200 billion. With five complete shuttles built and a total of 135 missions it works out to be about $1.5 billion per shuttle flight. The Space Shuttle *Endeavour* that was built to replace the *Challenger* space shuttle cost roughly $1.7 billion.[7] Cost per seat works out to be about $170 million in 2020 dollars.[8]

INTERNATIONAL SPACE STATION: $150 BILLION

The $150 billion is a shared cost between all participating nations. The U.S. invested almost $75 billion in the ISS which included construction, operating costs and transportation. The European share was around €8 billion that was spread over the entire almost 30-year program. This adds up to just one Euro spent by every European every year.[9] Maintaining and supporting the International Space Station costs an additional $3 to $4 billion every year.[10]

EMU SPACESUIT: $150 MILLION

The Extravehicular Mobility Unit, or EMU, is the spacesuit that most of us think of when we imagine an astronaut floating above Earth while outside the International Space Station. These were built in 1974 and each is estimated to have cost between $17 million and $22 million, which in today's money is close to $150 million.

These suits are really one-person spacecraft that have to keep the wearer alive in the vacuum of space. It also must allow them to maneuver and perform work on the space station for hours at a time. It protects them from the radiation of the sun, tiny meteorites flying at orbital speeds of 17,500 mph (28,000 kph), extreme temperatures around 248°F (120°C) in the sun and -148°F (-100°C) on the dark side of the Earth or Moon. It also carries enough oxygen to spend up to seven hours in space.[11]

18 suits were developed for the Space Shuttle program and they were originally built to last only 15 years. These suits are now over 45 years old, and there are only 11 suits left, in various conditions. Only four of these suits are flight-ready spacesuits. They are currently onboard the International Space Station.[12] Fortunately, NASA is working on a new generation of EMU spacesuits.

ARTEMIS XEMU SUIT
NASA has invested about $250 million to develop the new suits that will be used on the Moon, called the xEMU suit.[13] These suits have incorporated lessons learned from the Apollo suits and have made several enhancements that will improve mobility and they greatly increase the time astronauts are able to spend on the surface of the Moon.

EXPENSIVE GLOVES
While it's called a spacewalk, it should really be called a "handwalk" because there isn't any walking happening when astronauts are outside the space station on an EVA. To move around they use their hands to push and pull themselves using handrails, which is called translating.[14] They also need their hands to use a variety of tools as they work.

So protecting one's hands is crucial. Spacesuit gloves are one of the most expensive pieces on the spacesuit because they are one of the most complex parts of the suit. Among other factors, the gloves include a system of pulleys and strings that hold them together, and have a system of heat radiators to help keep hands warm.[15] The exact cost of the original spacesuit gloves is hard to track down, but NASA awarded $200,000 to the winner of their Astronaut Glove Challenge in 2007.[16]

MMU JETPACK: $26.7 MILLION
NASA invested $26.7 million into building the Manned Maneuvering Units (MMUs) that allow astronauts to fly untethered away from the space shuttle. The two operational flight units that flew to space

were valued at about $10 million each. Both units flew in February 1984 when astronauts Bruce McCandless II and Bob Stewart became the first "untethered" spacewalkers in history.[17]

 DID YOU KNOW?

The famous photo showing Bruce McCandless using the MMU was taken by Robert "Hoot" Gibson in 1984.[18] It's one of the most famous space photos ever taken and was on the cover of both *National Geographic* and *Aviation Week*.[19] *Photo courtesy of* NASA.

SPACE SHUTTLE TOILET: $30 MILLION

The space shuttle toilet that included his and hers funnels and a fan that would pull fluids and feces down where it should go cost $30 million.

ISS TOILET: $19 MILLION

The original toilet onboard the ISS on the American side cost $19 million and was purchased from the Russians. The Russian Zvezda module already had a toilet, but with the ISS accommodating up to six people at a time, a second toilet was badly needed, especially if the only one were to break down. This toilet can also recycle a large percentage of urine into filtered water.[20]

NEW ISS TOILET: $23 MILLION

Finally, a toilet that was built with females in mind, and it only cost $23 million. I'm sure if you ask the female astronauts, the new toilet is worth every dollar. This new toilet, also called the Universal Waste Management System, or UWMS, was sent to the ISS in November 2020. It has lots of new features and will also recycle 90% of urine into some of the cleanest water ever recycled.[21] The $23 million was a 2-for-1 deal, and the second will be going to the Moon onboard the Orion spacecraft.[22]

COST PER SEAT IN THE RUSSIAN SOYUZ

In 2008, when the space shuttle was still in use, Roscosmos charged NASA $21.8 million per Soyuz seat. The space shuttle was retired in 2011 and the Soyuz became the only way to send people to the International Space Station. In 2018, the cost per seat on the Soyuz was $81 million. In 2020, the price tag rose to $90.3 million. Since 2006, NASA has paid an average of $56.3 million per seat for the 71 planned and completed missions.[23]

SEAT ON SPACEX'S CREW DRAGON

A seat on SpaceX's Crew Dragon works out to be between $60 to $67 million. That makes it the least expensive cost per seat compared to previous spacecraft.[24]

COST PER SEAT ON BOEING'S STARLINER

NASA will pay between $91 and $99 million per seat onboard Boeing's Starliner spacecraft.[25]

NASA'S BUDGET & IMPACT

NASA budget for 2019 was $21.5 billion, which is less than half of 1% of the total federal budget. With that $21.5 billion, NASA generated more than $64 billion, supported 312,000 jobs nationwide, resulted in nearly $7 billion in federal, state and local tax revenues and impacted all 50 states.[26]

Since 1976, there have been over 2,000 spinoff technologies that have impacted Americans on a daily basis. Most recently, engineers at NASA's Jet Propulsion Lab developed a ventilator specifically for coronavirus patients in just 37 days. After receiving emergency use authorization from the Food and Drug Administration, the ventilator design was made available to manufacturers for free.[27]

The International Space Station costs each American less than $12 a year. For comparison, American adults spend an average of $1,200 on personal electronics each year. The cheapest Netflix plan is $8.99 a month.

ESA'S BUDGET

ESA's budget for 2021 is €6.49 billion. On average, every citizen of an ESA Member State pays in taxes about the same as the price of a movie ticket.[28]

NOTES

1. "Inflation Rate between 1969–2021," Inflation Calculator, accessed May 31, 2021, https://www.in2013dollars.com/us/inflation/1969?endYear=2020&amount=25000000000 .

2. Richard Hollingham, "Apollo in 50 Numbers: The Cost," BBC Future, July 12, 2019, https://www.bbc.com/future/article/20190712-apollo-in-50-numbers-the-cost.

3. Casey Dreier, "NASA's Commercial Crew Program Is a Fantastic Deal," The Planetary Society, May 19, 2020, https://www.planetary.org/articles/nasas-commercial-crew-is-a-great-deal-for-the-agency.

4. Hollingham, "Apollo in 50 Numbers."

5. "Mir Space Station," NASA, accessed May 31, 2021, https://history.nasa.gov/SP-4225/mir/mir.htm.

6. Claude Lafleur, "Costs of US Piloted Programs," *Space Review*, March 8, 2010, https://www.thespacereview.com/article/1579/1.

7. Brian Dunbar, "Space Shuttle and International Space Station," ed. Nancy Bray, NASA, August 3, 2017, https://www.nasa.gov/centers/kennedy/about/information/shuttle_faq.html#1.

8. Dreier, "NASA's Commercial Crew Program."

9. "How Much Does It Cost?" European Space Agency, accessed May 31, 2021, https://www.esa.int/Science_Exploration/Human_and_Robotic_Exploration/International_Space_Station/How_much_does_it_cost.

10. Office of Audits, "Extending the Operational Life of the International Space Station until 2024," Audit Report,, September 18, 2014, https://oig.nasa.gov/audits/reports/FY14/IG-14-031.pdf.

11. "The Space Shuttle Extravehicular Mobility Unit (EMU)," *Suited for Spacewalking* (EG-1998-03-112-HQ), 1998, https://www.nasa.gov/pdf/188963main_Extravehicular_Mobility_Unit.pdf.

12. Andy Ash, "Why NASA Spacesuits Are So Expensive," *Business Insider*, April 20, 2020, https://www.businessinsider.com/why-nasa-spacesuits-so-expensive-cost-russia-spacex-2020-3.

13. Ash, "Why NASA Spacesuits Are So Expensive."

14. Scott Kelly, *Endurance: A Year in Space, A Lifetime of Discovery* (New York: Alfred A. Knopf, 2017), 280.

15. Ash, "Why NASA Spacesuits Are So Expensive."

16. David Shiga, "New Spacesuit Glove Beats NASA's, Hands Down," *New Scientist*, May 4, 2007, https://www.newscientist.com/article/dn11794-new-spacesuit-glove-beats-nasas-hands-down/.

17. Ben Evans, "'To Face Their Wives': 30 Years Since First Untethered Spacewalk (Part 1)," AmericaSpace, February 1, 2014, https://www.americaspace.com/2014/02/01/to-face-their-wives-30-years-since-the-first-untethered-spacewalk-part-1/.

18. Anne Broache, "Footloose," Smithsonian Institution, August 1, 2005, https://www.smithsonianmag.com/science-nature/footloose-82112792/.

19. Evans, "To Face Their Wives."

20. Dhiram Shah, "The Most Expensive Toilet in the World Purchased by NASA," Newlaunches, July 9, 2007, https://www.newlaunches.com/archives/the_most_expensive_toilet_in_the_world_purchased_by_nasa.php.

21. Brian Dunbar, "Boldly Go! NASA's New Space Toilet," ed. Darcy Elburn, NASA, September 24, 2020, https://www.nasa.gov/feature/boldly-go-nasa-s-new-space-toilet-offers-more-comfort-improved-efficiency-for-deep-space/.

22. Loren Grush, "NASA Is about to Launch an Upgraded Microgravity Toilet to the International Space Station," *The Verge*, October 1, 2020, https://www.theverge.com/2020/10/1/21495881/nasa-microgravity-toilet-universal-waste-management-system-iss.

23. William Harwood, "NASA Uses Final Purchased Soyuz Seat for Wednesday Flight to Station," Spaceflight Now, October 13, 2020, https://spaceflightnow.com/2020/10/13/nasa-uses-final-purchased-soyuz-seat-for-wednesday-flight-to-station.

24. Dreier, "NASA's Commercial Crew Program."

25. Dreier, "NASA's Commercial Crew Program."

26. Robert Frost, "20 Uninterrupted Years of Human Occupation of Space," Quora, accessed May 31, 2021, https://qr.ae/pNbyW5.

27. Brian Dunbar, "NASA Report Details How Agency Significantly Benefits US Economy," ed. Karen Northon, NASA, January 4, 2021, https://www.nasa.gov/press-release/nasa-report-details-how-agency-significantly-benefits-us-economy.

28. "ESA Facts," European Space Agency, accessed May 31, 2021, http://www.esa.int/About_Us/Corporate_news/ESA_facts.

HOW NASA & SPACE EXPLORATION HAS BENEFITED US

A UNIQUE PERSPECTIVE

Astronauts and cosmonauts onboard the ISS have a unique perspective of our world and are able to see things that no one else can. They are able to help predict and prevent damage from flash flooding, hurricanes, earthquakes, tsunamis, volcanic eruptions and wildfires and also contribute to responses for these natural disasters.[1]

CELL PHONE CAMERAS, HI-DEF VIDEO & PILL CAMERAS

In the 1990s, a NASA engineer invented a sensor that requires very little power, is very small and can produce high-quality images. This technology is now in our cell phone cameras, webcams, in high-definition video and used for the "pill cameras" that allow for noninvasive endoscopy medical procedures. It's made it possible for NASA to capture distant galaxies, newly-discovered stars and astronomical events. We use this technology every day, thanks to NASA.[2]

SPECIAL SNOW GOGGLE FILTER

Have you ever tried to ski or snowboard on a bright sunny day? It's really hard to see with the light reflecting off the snow. This is partly caused by the blue light in sunlight. NASA designed a filter that blocks up to 95% of blue light. Snow goggles now have these blue-light cancellation filters that make it so skiers and snowboarders can see where they're going.[3]

WIRELESS HEADSETS

NASA worked with existing companies to improve the same technology that allowed them to communicate with their astronauts on the Moon.

These advancements led to more compact, comfortable and wireless headsets. These are used by pilots, professionals and even gamers.[4]

INFRARED EAR THERMOMETERS

The same technology astronomers use to measure the temperature of distant stars and planets is what we use to measure temperatures in humans with handheld infrared thermometers. NASA developed this infrared technology that can measure the energy emitted from the eardrum or forehead to record temperatures. So next time you have your temperature taken by an infrared temperature, you can thank NASA![5]

CORDLESS DRILLS AND HANDHELD VACCUUMS

When the astronauts went to the Moon it was important that they bring back lunar rock and soil samples. To do this, they needed a special drill that could extract core samples as much as 10 feet (3 m) below the lunar surface. NASA worked with the company Black and Decker to create a special, cordless drill for this very purpose. This later led to the cordless drills and handheld vacuums we use today.[6]

HIGHLY-EFFECTIVE WATER FILTERS

Every drop counts! So NASA uses state-of-the-art technology to filter water on the International Space Station because water is a precious resource in space. The same technology that recycles and filters sweat, urine and condensation into purified water has been used on Earth to create water filters for consumers. One such product is the ÖKO water bottle that filters water just by squeezing the bottle. This has been used for international travelers, camping, sports, biking and more. It's rated as effective in more than 120 countries.[7]

FREEZE-DRIED FOOD TECHNOLOGY

While the early astronauts and cosmonauts ate food from squeeze tubes, NASA needed to find a lightweight way to package food that would last for longer missions and didn't need refrigeration. They also wanted to provide more variety of foods that wouldn't or shouldn't fit in a squeeze tube. NASA partnered with several food companies and freeze-dried foods were invented.

To freeze-dry food, cooked food is rapidly frozen to -40°F (-40°C). This turns any water to ice crystals. The food is then evaporated in a vacuum that then removes any moisture. The vacuum-packed food doesn't require refrigeration, retains 98% of its nutritional value, is only 20% of its original weight and can last almost indefinitely. It also allows for a greater variety of food available. All you need to do is add water. This has not only been used by astronauts, but has become popular among campers, backpackers and as use as emergency food. This is also what makes astronaut ice cream, astronaut ice cream.[8]

ENRICHED BABY FOOD

Looking for ways to support long-duration space travel, NASA funded experiments using algae that could be used for food supply, oxygen generation and waste disposal. These experiments found a species of algae that produces omega-3 fatty acids that are found naturally in the body.

Almost all U.S. baby formulas are now enriched with these omega-3 fatty acids. Omega-3 fatty acid is believed to help with mental and visual development. These additives aren't just found in baby formula, they can be found in other food for children, adults, pets and livestock.[9]

TELEMEDICINE

NASA has had a long history with telemedicine and has played a significant part in developing it into what it is today. Over the years, NASA has used forms of telemedicine to monitor the health of astronauts in space and this technology has been helpful worldwide during the coronavirus pandemic. It's made it possible to prevent potential overcrowding, to see patients during times of lockdown and to reduce the transmission of the virus among potentially vulnerable people.[10]

MEMORY FOAM

To improve the safety and comfort of commercial air travel in the 1960s, a NASA project developed temper foam, or what's more commonly known as Memory Foam. Memory Foam is soft yet absorbs impacts. Memory Foam has been used in mattresses, pillows, hospital beds, footwear, prosthetics, motorcycle saddles and so much more. It is also still widely used for its original purpose: commercial and military airplane seats.[11]

FIREPROOF MATERIAL

After the fire on Apollo1 that killed all three astronauts onboard, NASA went to work to create something fireproof to protect the astronauts. They found a heat-resistant material called polybenzimidazole, or PBI. Firefighter suits now contain PBI and it can withstand heat up to 1,300°F (704°C).[12]

KEEPING FOOD & US SAFE

The last thing NASA wants is for astronauts to get food poisoning while in space, so they were extra cautious when preparing food for astronauts in the early space program. NASA soon realized that the food industry's standard of randomly testing food at the end of the process wasn't good enough. NASA and Pillsbury[13] came up with the Hazard Analysis and Critical Control Point, or HAACP. This method is now used all over the world for food production, and even pharmaceuticals.[14]

OLYMPIC-MEDAL WINNING SWIMSUIT

Speedo used NASA research from their wind-tunnel testing to develop their LZR Racer swimsuit also called "the rubber suit." Speedo engineers learned which materials and seams worked best to reduce drag during swimming. They created a suit that "compressed a swimmer's body into a streamlined tube and trapped air, adding buoyancy and reducing drag."[15]

The full-body LZR Racer swimsuit first appeared at the 2008 Beijing Olympics and 98% of the medal winners and world-record breaker swimmers wore this suit. These suits are now banned from international competitions, but a modified version is still available and is popular among professional swimmers.[16]

NOTES

1. Brian Dunbar, "Clear High-Definition Images Aid Disaster Response," ed. Michael Johnson, NASA, November 12, 2019, https://www.nasa.gov/mission_pages/station/research/news/b4h-3rd/eo-clear-high-def-images.

2. "Digital Cameras," Home and City, NASA, accessed May 31, 2021, https://homeandcity.nasa.gov/nasa/livingroom/164/digital-cameras.

3. "Blue-Light Cancellation," Home and City, NASA, accessed May 31, 2021, https://homeandcity.nasa.gov/nasa/livingroom/165/bluelight-cancellation.

4. "Wireless Headset," Home and City, NASA, accessed May 31, 2021, https://homeandcity.nasa.gov/nasa/livingroom/171/wireless-headset.

5. "Infrared Ear Thermometers," Home and City, NASA, accessed May 31, 2021, https://homeandcity.nasa.gov/nasa/bathroom/233/infrared-ear-thermometers.

6. "Portable Cordless Vacuums," Home and City, NASA, accessed May 31, 2021, https://homeandcity.nasa.gov/nasa/kitchen/158/portable-cordless-vacuums.

7. "Enhanced Water Bottles Filter Water on the Go," NASA, accessed May 31, 2021, https://spinoff.nasa.gov/Spinoff2013/cg_1.html.

8. "Freeze Dried Technology," Home and City, NASA, accessed May 31, 2021, https://homeandcity.nasa.gov/nasa/kitchen/163/freeze-dried-technology.

9. "Enriched Baby Food," Home and City, NASA, accessed May 31, 2021, https://homeandcity.nasa.gov/nasa/kitchen/162/enriched-baby-food.

10. Brian Dunbar, "NASA and Telemedicine," ed. Thalia Patrinos, NASA, April 7, 2020, https://www.nasa.gov/feature/nasa-and-telemedicine.

11. "Forty-Year-Old Foam Springs Back with New Benefits," NASA, accessed May 31, 2021, https://spinoff.nasa.gov/Spinoff2005/ch_6.html.

12. Jacob Margolis and Christopher Intagliata, "Space Spinoffs: The Technology to Reach the Moon Was put to Use Back on Earth," NPR, July 20, 2019, https://www.npr.org/2019/07/20/742379987/space-spinoffs-the-technology-to-reach-the-moon-was-put-to-use-back-on-earth.

13. "Food Safety Program for Space Has Taken over on Earth," NASA Spinoff 2021, December 2020, https://spinoff.nasa.gov/sites/default/files/2020-12/NASA_Spinoff-2021.pdf.

14. "Food Safety Systems," Home and City, NASA, accessed May 31, 2021, https://homeandcity.nasa.gov/nasa/grocery/205/food-safety-systems.

15. Jim Morrison, "How Speedo Created a Record-Breaking Swimsuit," Scientific American, July 27, 2012, https://www.scientificamerican.com/article/how-speedo-created-swimsuit/.

16. "Olympic Swimsuit," Home and City, NASA, accessed May 31, 2021, https://homeandcity.nasa.gov/nasa/sports/267/olympic-swimsuit.

QUOTES FROM ASTRONAUTS

"I thought at one point, if you could be up in heaven, this is how you would see the planet. And then I dwelled on that and said, no, it's more beautiful than that. This is what heaven must look like. I think of our planet as a paradise. We are very lucky to be here."
– Mike Massimino, NASA astronaut.

"Space exploration is for me a great adventure of the human spirit, a shared experience that nourishes the noblest part of us, elevating us above meanness and boredom."
– Samantha Cristoforetti, first Italian woman in space.

VALENTINA TERESHKOVA, FIRST WOMAN IN SPACE
"Once you've been in space, you appreciate how small and fragile Earth is."

"Anyone who has spent any time in space will love it for the rest of their lives. I achieved my childhood dream of the sky."

"It [the Earth] was breathtakingly beautiful, like something out of a fairy tale. There is no way to describe the joy of seeing Earth. It is blue, and more beautiful than any other planet."

SALLY RIDE, THE FIRST AMERICAN WOMAN IN SPACE
"The stars don't look bigger, but they do look brighter."

"All adventures, especially into new territory, are scary."

"Weightlessness is a great equalizer."

MAE JEMISON, THE FIRST AFRICAN AMERICAN WOMAN IN SPACE

"Never limit yourself because of others' limited imagination; never limit others because of your own limited imagination."

"Once I got into space, I was feeling very comfortable in the universe. I felt like I had a right to be anywhere in this universe, that I belonged here as much as any speck of stardust, any comet, any planet."

QUOTES FROM APOLLO ASTRONAUTS

"It suddenly struck me that that tiny pea, pretty and blue, was the Earth. I put up my thumb and shut one eye, and my thumb blotted out the planet Earth. I didn't feel like a giant. I felt very, very small."
– **Neil Armstrong, the first person to step on the Moon.**

"The view of the Earth from the Moon fascinated me—a small disk, 240,000 miles away...Raging nationalistic interests, famines, wars, pestilence don't show from that distance."
– **Frank Borman, Apollo 8 commander, one of the first men to travel to the Moon.**

"We came all this way to explore the Moon, and the most important thing is that we discovered the Earth."
– **William Anders, Apollo 8 astronaut, one of the first men to travel to the Moon.**

"Now I know why I'm here. Not for a closer look at the Moon, but to look back at our home, the Earth."
– **Alfred Worden, Apollo 15 astronaut, one of only 24 men to fly to the Moon.**

QUIZ YOURSELF

1. Who was the first human in space?
   A. John Glenn
   B. Valentina Tereshkova
   C. Yuri Gagarin
   D. Alan Shepard

2. Astronauts must know how to swim.
   A. True
   B. False

3. How many people have walked on the Moon?
   A. 24
   B. 2
   C. 12
   D. 10

4. What does the "naut" part of astronaut and cosmonaut mean?
   A. Explorer
   B. Pioneer
   C. Sailor
   D. Traveler

5. Who said, "One small step for man, one giant leap for mankind?"
   A. Buzz Aldrin
   B. Neil Armstrong
   C. John Glenn
   D. Jim Lovell

6. Who was the first woman in space?
   A. Valentina Tereshkova
   B. Sally Ride
   C. Svetlana Savitskaya
   D. Mae Jemison

7. The crew of Apollo 11 left 106 objects on the Moon.
   A. True
   B. False

8. How many American flags were left on the Moon and are still there?
   A. 0
   B. 4
   C. 6
   D. 8

9. Who was the first space tourist to pay his own way to the International Space Station?
   A. Elon Musk
   B. Richard Garriott
   C. Richard Branson
   D. Dennis Tito

10. What was the name of Apollo 11's lunar module that landed on the Moon?
   A. Snoopy
   B. Columbia
   C. Eagle
   D. Falcon

11. During the Space Flight Participant Program, NASA officials discussed having Big Bird from the TV show Sesame Street fly onboard the space shuttle.
   A. True
   B. False

12. Russia had their own version of the space shuttle.
   A. True
   B. False

13. What were the names of the six space shuttles?
   A. Columbia, Eagle, Endurance, Intrepid, Perseverance and Opportunity
   B. Enterprise, Columbia, Challenger, Discovery, Atlantis and Endeavour
   C. Curiosity, Spirt, Sojourner, Challenger, Endurance and Discovery
   D. Challenger, Columbia, Discovery, Ingenuity, Tenacity and Courage

14. According to the International Astronautical Federation, the beginning of space begins at the Kármán line. How high is that?
   A. 100 miles or 160 km
   B. 55 miles or 88 km
   C. 62 miles or 100 km
   D. 93 miles or 150 km

15. What is the name of the current Russian spacecraft that ferries cosmonauts to the International Space station and back?
   A. Soyuz
   B. Vostok
   C. Buran
   D. Voskhod

16. Who was the third Apollo 11 crew member who stayed onboard the Command Module while the other two men walked on the Moon?
A. Charlie Duke
B. Jim Lovell
C. Ken Mattingly
D. Michael Collins

17. Which of the following are NOT items that came as a result of space exploration?
A. Cell phone cameras
B. Velcro
C. Memory Foam
D. Portable Cordless Vacuums

18. Astronauts are scuba certified.
A. True
B. False

19. Crews from Apollo 11, 12 and 14 were required to quarantine for three weeks after returning from the Moon in case they brought back moon germs.
A. True
B. False

20. What does NASA stand for?
A. New American Space Agency
B. National Aeronautics and Space Administration
C. National Air and Space Agency
D. Navy and Airforce Space Administration

QUIZ ANSWERS

1. Who was the first human in space?
A. John Glenn
B. Valentina Tereshkova
C. Yuri Gagarin
D. Alan Shepard

2. Astronauts must know how to swim.
A. True
B. False

3. How many people have walked on the Moon?
A. 24
B. 2
C. 12
D. 10

4. What does the "naut" part of astronaut and cosmonaut mean?
A. Explorer
B. Pioneer
C. Sailor
D. Traveler

5. Who said, "One small step for man, one giant leap for mankind?"
A. Buzz Aldrin
B. Neil Armstrong
C. John Glenn
D. Jim Lovell

6. Who was the first woman in space?
A. Valentina Tereshkova
B. Sally Ride
C. Svetlana Savitskaya
D. Mae Jemison

7. The crew of Apollo 11 left 106 objects on the Moon.
A. True
B. False

8. How many American flags were left on the Moon and are still there?
A. 0
B. 4
C. 6
D. 8

9. Who was the first space tourist to pay his own way to the International Space Station?
A. Elon Musk
B. Richard Garriott
C. Richard Branson
D. Dennis Tito

10. What was the name of Apollo 11's lunar module that landed on the Moon?
A. Snoopy
B. Columbia
C. Eagle
D. Falcon

11. During the Space Flight Participant Program, NASA officials discussed having Big Bird from the TV show *Sesame Street* fly onboard the space shuttle.
A. True
B. False

12. Russia had their own version of the space shuttle.
A. True
B. False

13. What were the names of the six space shuttles?

A. Columbia, Eagle, Endurance, Intrepid, Perseverance and Opportunity

B. Enterprise, Columbia, Challenger, Discovery, Atlantis and Endeavour

C. Curiosity, Spirt, Sojourner, Challenger, Endurance and Discovery

D. Challenger, Columbia, Discovery, Ingenuity, Tenacity and Courage

14. According to the International Astronautical Federation, the beginning of space begins at the Kármán line. How high is that?

A. 100 miles or 160 km

B. 55 miles or 88 km

C. 62 miles or 100 km

D. 93 miles or 150 km

15. What is the name of the current Russian spacecraft that ferries cosmonauts to the International Space station and back?

A. Soyuz

B. Vostok

C. Buran

D. Voskhod

16. Who was the third Apollo 11 crew member who stayed onboard the Command Module while the other two men walked on the Moon?

A. Charlie Duke

B. Jim Lovell

C. Ken Mattingly

D. Michael Collins

17. Which of the following are NOT items that came as a result of space exploration?

A. Cell phone cameras

B. Velcro

C. Memory Foam

D. Portable Cordless Vacuums

18. Astronauts are scuba certified.

A. True

B. False

19. Crews from Apollo 11, 12 and 14 were required to quarantine for three weeks after returning from the Moon in case they brought back moon germs.

A. True

B. False

20. What does NASA stand for?

A. New American Space Agency

B. National Aeronautics and Space Administration

C. National Air and Space Agency

D. Navy and Airforce Space A. Administration

GLOSSARY OF TERMS & ACRONYMS

ARED (Advanced Resistance Exercise Device)
The name of the weightless weightlifting machine astronauts use while on the space station to help counteract the bone density and muscle mass loss that occurs from living in microgravity.

AQUARIUS
The world's only undersea research station, laboratory and habitat used in NEEMO. Aquarius is where NASA sends groups of astronauts, engineers and scientists to live and train for up to three weeks to prepare them for future space missions.

Located 3.5 miles (5.6 km) off Key Largo, Florida in the Florida Keys, it sits on the ocean floor 62 feet (19 m) below the surface.

CAPCOM
Shortened acronym for Capsule Communicator. The role in Mission Control normally filled by an astronaut, who is responsible for talking to the astronauts in space.

CAVES
Acronym for Cooperative Adventure for Valuing and Exercising Human Behavior and Performance Skills. This is an ESA astronaut training course in cave environments that takes place over several days. It simulates a space-like environment where astronauts carry out exploration activities that help them develop and practice efficiency, cooperation and risk management with their crew members.

CEVIS (Cycle Ergometer with Vibration Isolation and Stabilization)
This is the stationary bicycle with no seat that the astronauts use for cardiovascular training on the ISS.

CSA
Acronym for Canadian Space Agency. The Canadian equivalent of NASA, ESA, JAXA and Roscosmos.

CO2
Carbon Dioxide. This is what we breathe out and what needs

to be removed from spacesuits and the space station.

CSL
Acronym for Crew Support Laptops. These are one of two laptops kept in the crew's quarters onboard the ISS. It's the computer that allows them to connect to a computer in Houston remotely to then access the internet.

CYGNUS
One of the resupply cargo spacecrafts that flies to the ISS. It was developed by the American company Northrop Grumman Innovation Systems, formally Orbital ATK. The Cygnus cargo ships have been flying on behalf of NASA since 2013.

DPC (Daily Planning Conference)
Each workday on the ISS, there are two DPCs, one in the morning and one in the evening. It's a meeting held via radio between the ISS astronauts and cosmonauts, and mission control centers around the world.

DRAGON
SpaceX's spacecraft that comes in two varieties: Crew Dragon and Cargo Dragon. It has carried cargo to the ISS on behalf of NASA since 2012. It has launched crews to the ISS since May 2020.

EMU (Extravehicular Mobility Unit)
This is the pressurized American spacesuit astronauts use while performing spacewalks.

ERA
Acronym for the European Robotic Arm developed by ESA for the International Space Station.

ESA
Acronym for European Space Agency. The European equivalent of NASA, CSA, JAXA and Roscosmos.

EUROCOM
This is the name of the CAPCOM position at the Columbus Control Center or Col-CC responsible for speaking with the ISS astronauts on the Space-to-Ground radio, usually about activities in the ESA Columbus module.

EVA (Extravehicular Activity)
This is what most people call a spacewalk while wearing a pressurized spacesuit.

FLUID SHIFT

The shifting and redistribution of bodily fluids caused by weightlessness.

FOD (Foreign Object Debris)

Once spaceships enter microgravity, FOD come out of hiding and make its appearance. FOD can be tiny nuts and bolts, metal shavings, staples, hair, dust or plastic flotsam. There are people at the Kennedy Space Center whose entire job is to keep the spacecrafts clean and FOD-free.

G

G used in this book refers to a unit of measurement for the perceived weight felt in a spacecraft. For reference, 1 G is equal to a person's or object's normal weight on the Earth's surface.

GLAVNI

The name of the CAPCOM position at the Tsentr Upravleniya Poloyotami (TsUP), or mission control center, near Moscow, Russia responsible for speaking with the ISS astronauts on Space-to-Ground radio.

ISS

Acronym for the International Space Station.

JAXA

Acronym for Japanese Space Agency. The Japanese equivalent of NASA, ESA, CSA and Roscosmos.

J-COM

The name of the CAPCOM position at the Control Center of the Japanese Space Agency responsible for speaking with the ISS astronauts on Space-to-Ground radio usually about activities in the Japanese module JEM.

JEM

Acronym for the Japanese Experiment Module on the ISS.

MAG

NASA's acronym for Maximum Absorbency Garment or adult diapers the astronauts use during launch, re-entry and during spacewalks.

MECO

Abbreviation for Main Engine Cutoff and pronounced as "mee-ko."

MICROGRAVITY

The condition where people and objects appear to be weightless and float in space because they are in free fall. Microgravity doesn't mean zero gravity or the absence of gravity. It means there is a small amount of gravity, but in orbit, everything is in free fall and falling at the same rate. Since everything is all falling together, everything appears to float when compared with the spacecraft.

MMU

Acronym for Manned Maneuvering Unit. The MMU was an earlier jet-pack before the SAFER jet-pack. It contained enough propellant for a six-hour EVA depending on how much maneuvering is done.

NASA

Acronym for National Aeronautics and Space Administration, the American space agency. The American equivalent to CSA, ESA, JAXA and Roscosmos.

NBL (Neutral Buoyancy Laboratory)

This is NASA's training facility with a giant pool where astronauts wear the EMU suit and train for EVAs. At the bottom are real-size models of the non-Russian modules of the ISS.

NEEMO (NASA Extreme Environment Mission Operations)

This is a NASA program in the Florida Keys where astronauts, engineers and scientists are sent to an underwater laboratory called Aquarius for missions up to three weeks. They conduct experiments, test technology and simulate how things might work in space.

ORLAN

This is the name of Russian's pressurized spacesuit used for EVAs. Orlan in Russian means "sea eagle."

PAYCOM

The name for the CAPCOM position at the Payload Operations Center at NASA's Marshal Space Flight Center in Huntsville, Alabama that is responsible for speaking with the ISS astronauts on Space-to-Ground radio, usually about NASA experiments.

PLSS

Acronym for Portable Life Support System or backpack that incorporates a small jet-pack that is worn as part of an astronaut's spacesuit.

PROGRESS

Russia's cargo vehicle and the twin of the Soyuz.

RMS

Acronym for Remote Manipulator System which

are robotic arms like the Canadarm2 or Dextre, that are used to capture and release satellites, free-floating cargo spacecrafts and help maneuver spacewalking astronauts.

ROSCOSMOS

The name of the Russian space agency. The Russian equivalent of NASA, CSA, ESA and JAXA.

SAFER

Acronym for Simplified Aid for EVA Rescue. This small jet-pack is incorporated into the PLSS, which is a new and improved version of the MMU. This jet-pack relies on small nitrogen-jet thrusters that allow an astronaut to maneuver in space in case they become detached from the space station during a spacewalk.

The SAFER jet-pack comes with 24 high-pressure thrusters that give control in six axes – pitch, roll, yaw, forwards/backwards, sideways and up/down.

SHLEMOFON

The cloth headgear with integrated earphone and microphone for radio communication worn under the helmet of the Sokol suit in the Soyuz spacecraft.

SNOOPY CAP

The American equivalent to the Shlemofon. It's the cloth headgear with the integrated earphone and microphone used for radio communication and worn under the EMU suit helmet. The name comes from the aviator cap that Snoopy wears in the Peanuts cartoons by Charles Schulz.

SOKOL

Pressure suit worn by astronauts flying in the Russian spacecraft Soyuz. An emergency suit, it is designed to sustain an astronaut's life in the event of atmospheric loss in the cabin and cannot be used in space outside a spacecraft.

SOYUZ

Soyuz is the name of the Russian spacecraft, rocket and space program and is the word for union in Russian.

SPACE-TO-GROUND

The radio communication channel used by astronauts and cosmonauts to speak from the ISS to Mission Control Centers around the world.

SPACE SHUTTLE ORBITER

The Space Shuttle Orbiter was NASA's partially reusable

spaceplane built to shuttle astronauts to and from space and was able to orbit the Earth. Five orbiters were built during the Space Shuttle program. It was about the same size and weight as a DC-9 aircraft and could normally carry up to seven crew members.

STS

Space Transportation System. Space Shuttle missions were named STS and then a number. The first NASA space shuttle orbital spaceflight was STS-1. The last Space Shuttle mission was STS-135.

TSUP

The acronym for the Roscomos Russian Mission Control Center in Korolev in the larger Moscow metropolitan area. It's pronounced like "soup."

TVIS

Acronym for Treadmill with Vibration Isolation Stabilization System. This is the special treadmill astronauts use while living on the International Space Station. It uses a special harness similar to a backpacker's harness that allows them to adjust the weight load. The harness is attached to the treadmill by two bungees which is what holds the astronaut down and makes it possible for them to run on the treadmill.

UKSA

Acronym for United Kingdom Space Agency.

WEIGH-OUT

The process of attaching weights and floaters to the Orlan or EMU spacesuits to achieve neutral buoyancy when training underwater.

WHC

Acronym for Waste Hygiene Compartment which is the fancy name for the space toilet.

WHERE TO LEARN MORE

Find all of these links and more at: KnowledgeNuggetBooks.com/
Resources

NASA
The official NASA website is at: https://www.nasa.gov/

CANADA SPACE AGENCY
The Canadian Space Agency website has a wealth of video and
information about their astronauts and what it's like to live in space.
Visit their website at: https://www.asc-csa.gc.ca/eng/astronauts/
default.asp

EUROPEAN SPACE AGENCY
Want to learn about the European Space Agency? Want to see who
becomes the next ESA astronauts? They just recently opened up
applications. Find out the latest on their website: https://www.esa.
int/

ROSCOSMOS – RUSSIAN SPACE AGENCY
Learn about the Russian space program and the cosmonauts who
have made history on the English version of their official website
here: http://en.roscosmos.ru/

JAXA – JAPAN AEROSPACE EXPLORATION AGENCY
Learn more about Japan's Space Agency, the research they are doing
on the ISS, their launch vehicles and so much more. Visit their
official website here: https://global.jaxa.jp/

ASTRONAUT BIOGRAPHIES
Get all the information you could ever want about NASA astronauts
here: https://www.nasa.gov/astronauts

Find out the latest on the ESA astronauts here:
http://www.esa.int/Science_Exploration/Human_and_Robotic_
Exploration/Astronauts/European_astronauts

Learn about the active and former CSA astronauts here:
https://www.asc-csa.gc.ca/eng/astronauts/canadian/default.asp

NASA LIVE YouTube Streams
NASA has two different livestreams on YouTube, plus a host of
other playlists and videos to see what the astronauts are up to, the
Perseverance Rover on Mars and so much more. Find their official
YouTube channel here: https://www.youtube.com/channel/UCLA_
DiR1FfKNvjuUpBHmylQ

#AskNASA
Have a question for NASA? Use the hashtag #AskNASA on Twitter or
Instagram to ask your question and NASA will answer.

HOW TO PREPARE TO BECOME AN ASTRONAUT
NASA answers this here: https://www.nasa.gov/feature/10-ways-
students-can-prepare-to-beanastronaut

The CSA has lots of useful information about what it takes to be
a CSA astronaut found here: https://www.asc-csa.gc.ca/eng/
astronauts/how-to-become-an-astronaut/default.asp

ESA recently opened up applications to become an ESA astronaut.
Find out what qualities they look for:
https://www.esa.int/Science_Exploration/Human_and_Robotic_
Exploration/Astronauts/How_to_become_an_astronaut

The Japanese Space Agency answers this question here: https://
global.jaxa.jp/article/special/astronaut/yanagawa_e.html

ARTEMIS – SENDING HUMANS TO THE MOON AGAIN
Learn about what NASA is doing to send humans to the Moon again
here: https://www.nasa.gov/specials/artemis/

MARS
Want the latest information on what NASA is doing on Mars with the *Perseverance* Rover and *Ingenuity*, the Mars helicopter? Get all the latest info here: https://mars.nasa.gov/mars2020/

HOW SPACE EXPLORATION BENEFITS US
Want to learn about all the benefits space exploration has had on our lives here on Earth? Learn more here: https://homeandcity.nasa.gov/

VIEW ALL OF THE EXPERIMENTS PERFORMED ON THE ISS
To see and search for all of the experiments that have happened and are happening onboard the space station, visit: https://www.nasa.gov/mission_pages/station/research/experiments/explorer/index.html

QUORA
Quora is a great place to ask questions and find answers about a variety of things, but especially about astronauts and space exploration. There are even a few astronauts and folks who work at NASA who have answered several astronaut-related questions for others that you can read.

PHOTOS AND VIDEOS
Some of the best and most complete collection of photographs taken of the Earth since 1961 can be found and searched for here: https://eol.jsc.nasa.gov/

LIVE VIDEO FROM THE ISS
The live video from the International Space Station that includes views of the crew when they're on-duty, audio conversations between the crew and Mission Control and views of the earth from space at other times can be viewed here: https://www.nasa.gov/multimedia/nasatv/iss_ustream.html

VIEW EARTH LIVE FROM THE INTERNATIONAL SPACE STATION
A camera mounted on the ISS streams live video footage and shows the earth below as it flies by at 5 miles per second or 8 km per second.

Here's where you can see the view of the Earth as seen from the International Space Station: https://www.asc-csa.gc.ca/eng/iss/watch-live.asp

COMMUNICATING WITH THE ISS VIA HAM RADIO
AMATEUR RADIO ON THE INTERNATIONAL SPACE STATION PROGRAM
Learn how it all started, how you can apply to have your group speak to the crew onboard the ISS and what frequencies and call signs to use. Find that all here: https://www.ariss.org/

ASTRONAUTS AND THEIR CALL SIGNS
Get the full list of call signs for astronauts who are, or were at one time, licensed ham radio operators. See it here: https://www.ariss.org/hams-in-space.html

HOW TO HEAR THE INTERNATIONAL SPACE STATION
Find out what equipment you need to hear the ISS, how you can listen online, when to listen and what you can expect to hear at https://amsat-uk.org/beginners/how-to-hear-the-iss/

RECOMMENDED ASTRONAUT BOOKS
Books Written *By* Astronauts:
An Astronaut's Guide to Life on Earth: What Going to Space Taught Me About Ingenuity, Determination, and Being Prepared for Anything by Chris Hadfield

Ask an Astronaut: My Guide to Life in Space by Tim Peake

Diary of an Apprentice Astronaut by Samantha Cristoforetti

Endurance: My Year in Space. A Lifetime of Discovery by Scott Kelly

Handprints on Hubble: An Astronaut's Story of Invention by Kathryn Sullivan

How to Astronaut: An Insider's Guide to Leaving Planet Earth by Terry Virts

Go For Orbit: One of America's First Women Astronauts Finds Her Space by Rhea Seddon

Through the Glass Ceiling to the Stars: The Story of the First American Woman to Command a Space Mission by Col. Eileen M. Collins

Books Written *About* Astronauts:
Almost Astronauts: 13 Women Who Dared to Dream by Tanya Lee Stone

A Man on the Moon: The Voyages of the Apollo Astronauts by Andrew Chaikin

Fallen Astronauts: Heroes Who Died Reaching for the Moon by Colin Burgess and Kate Doolan

The Mercury 13: The Untold Story of Thirteen American Women and the Dream of Space Flight by Martha Ackermann

Packing for Mars: The Curious Science Of Life In Space by Mary Roach

RECOMMENDED ASTRONAUT WEBSITES
First British Astronaut, Helen Sharman
  Read her fascinating FAQs and learn how you can book Helen to come and speak to your group. Details at https://www. helensharman.uk/frequently-asked-questions/

First Canadian to Walk in Space, Chris Hadfield
Learn about Colonel Hadfield, find out about his latest book, learn about events in your area and sign up to get the latest at: https://chrishadfield.ca/

First Female Pilot and Female Commander of a space shuttle, Eileen M. Collins
Find out more about former astronaut and retired U.S. Air Force colonel, Eileen Collins. Learn more about her new book and read what she says about what it takes to become an astronaut. Details found on her official website: http://marklarson.com/eileencollins/one.html

First American Woman to Walk in Space, Kathy Sullivan
Sullivan has been to space and the bottom of the ocean. Read her blog and learn about her podcast where she talks about her historic experiences: http://kathysullivanastronaut.com/

RECOMMENDED ASTRONAUT-RELATED PODCASTS
ESA Explores
Get behind the scenes with the European Space Agency and find out how they train their astronauts, the science behind the ISS and what it will take to go to Mars.

Houston, We Have a Podcast
The official podcast of the Johnson Space Center in Houston, Texas where the host talks about all things space and how it works to get humans there. Lots of interviews with guests that talk spacesuits, artificial gravity and what it will take to get humans back to the Moon and to Mars.

Kathy Sullivan Explores
Astronaut Kathy Sullivan is the only person to have both walked in space and visited the deepest point in the ocean. In her new podcast she shares unique insights and perspectives about her experiences in space and in the ocean. She is also joined by a variety of guests which always leads to fascinating discussions.

Space and Things
A weekly podcast that explores all things about space. Learn about spacesuits, top tips for visiting Space Center Houston, how to talk to someone in space via Ham Radio and so much more. It's presented by historian Emily Carney and space nerd Dave Giles. More details at: https://www.spaceandthingspodcast.com

BBC World Service – Space
Listen to a collection of radio documentary programs that talk about all things space. Episodes cover what happen the day the Skylab space station fell to Earth, what it will take to get a woman on the Moon and the future of space flight. Get all of the episodes here: https://www.bbc.co.uk/programmes/p03bv899/episodes/player

MECO
MECO stands for Main Engine Cut Off and is a podcast by Anthony Colangelo. With over 200 episodes, you'll find topics, news and opinions about space tourism, Starship news, China's space station and so much more. Find it here: https://mainenginecutoff.com/podcast

Are We There Yet?
Listen to weekly space news updates and interviews with astronauts, engineers and visionaries by space reporter Brendan Byrne. Find it on Spotify and Apple Podcasts.

LEARN SOMETHING? PLEASE LEAVE A REVIEW

If you enjoyed this book, please share the word with others by leaving a **REVIEW.**

It helps them, helps me and you get to do a good deed for the day, which helps you.

You can leave a **REVIEW** for the book at the retailer of your choice with this easy link:

https://EverythingAboutAstronauts.com/Review

Thank you so very much!

DON'T FORGET YOUR BONUS

As a **special bonus** and as a **thank you** for purchasing this book, I'm giving you **early access and a free download** to the upcoming *Everything About Astronauts Volume 2.*

Volume 2 covers intriguing topics about what it's like to live and work in space, what foods astronauts

It's all FREE.

Download your bonus ebook chapter of *Everything About Astronauts Volume 2* here:

http://bit.ly/astronaut-bonus-chapter

or

SCAN ME

Enjoy!

INDEX

ABOUT THE AUTHOR

Marianne Jennings is a self-proclaimed "adventure craver" – she goes on crazy adventures with people she's never met, isn't afraid to try new things (like learning to sail in the middle of the Atlantic Ocean, despite being prone to seasickness), and proudly holds the title of Favorite Aunt to her 10 nieces and nephews. Marianne is a lover of new foods and new experiences, and wants to be remembered for being kind and generous.

Aside from embarking on all sorts of adventures, she is an accomplished international traveler and loves facts and trivia like astronauts love Tang. To help introduce other places, people, and cultures to others, she likes to share interesting and fun facts that are entertaining and memorable. Her popular, award-winning breakout title *So You Think You Know Canada, Eh?* kicked off her Knowledge Nugget Books series.

If you'd like to learn more or join her mailing list, you can connect with Marianne at https://knowledgenuggetbooks.com or on Instagram @ knowledgenuggetbooks.

ALSO BY MARIANNE JENNINGS

2020 Readers' Favorite International Gold Medal Winner

So You Think You Know CANADA, Eh?
Fascinating Fun Facts and Trivia about Canada for the Entire Family

Available as an e-book, paperback and audiobook.

Find it wherever fine books are sold - online or in stores.

Printed in Great Britain
by Amazon

12679478R00149